T0341040

Cumin (*Cuminum cyminum*)
Production and Processing

Editors

M. Kafi
M.H. Rashed Mohassel
A. Koocheki
M. Nassiri
Centre of Exellence for Special Crops
Faculty of Agriculture
Ferdowsi University of Mashhad
Iran

Science Publishers

Enfield (NH)   Jersey   Plymouth

CIP data will be provided on request.

SCIENCE PUBLISHERS
An Imprint of Edenbridge Ltd., British Isles.
Post Office Box 699
Enfield, New Hampshire 03784
United States of America

Website: *http://www.scipub.net*

sales@scipub.net (marketing department)
editor@scipub.net (editorial department)
info@scipub.net (for all other enquiries)

ISBN 1-57808-504-7 [10 digits]
　　　978-1-57808-504-0 [13 digits]

Published by Science Publishers, Enfield, NH, USA
An Imprint of Edenbridge Ltd.
Printed in India

Preface

A study of the history of agriculture reveals the success of Asian farmers in domesticating many plants and animals. These farmers have long been farming in an environment with scarcity of resources, particularly water. As a result their practices were based on optimum water utilization through appropriate cropping patterns and enhancement of water use efficiency. Utilization of underground water through a unique system of water extraction called 'qanat' or underground conduits is documented as a technology developed by Iranians in dry areas. Development of cropping systems based on low water requirement for plants such as cumin is another example of agricultural expertise of farmers in these areas. However, with the advent of new farming technologies in the last century based on genetic improvement, water availability and high input of chemical fertilizers and pesticides, indigenous technologies were sidelined and the focus shifted to new crops such as maize, soybean, potato, etc. and local crop species were not able to compete economically with such high yielding species and were consequently neglected. Most of these neglected crops are in the developing countries and not much support has been given by the international research organizations in terms of funding. This was also the case for medicinal plants, such as cumin, etc.

Against the above backdrop, the Center of Excellence for Special Crops was established in the Department of Agronomy and Plant

Breeding, Faculty of Agriculture, Ferdowsi University of Mashhad, Iran with a mandate to conduct research on such crops. One of the first attempts of the Center was to collect and document all the local and international literature on cumin, the main spice crop of Iran, India and many other countries. This book is based mainly on the results of research conducted locally. These findings have been analyzed comprehensively and compared with results from global sources.

This book contains ten chapters. In the first chapter, history, importance, acreage, production and utilization of cumin have been reviewed in general. In the second chapter, botany and plant characteristics are discussed, and in the third chapter ecophysiological criteria of cumin is analyzed. Production technology is elaborated in the fourth chapter and water requirement is the topic of chapter five. Pests and diseases of cumin are reviewed in the sixth chapter, and genetic and breeding aspects are detailed in chapter seven. Economic aspects of cumin are the main area of chapter eight, and, in chapter nine, processing, chemical composition and standards are described. Finally, research strategies are enumerated in chapter ten.

The editors would like to place on record their appreciation to the Center of Excellence for Special Crops, Department of Agronomy and Plant Breeding, Faculty of Agriculture, Ferdowsi University of Mashhad, Iran, for providing financial support which enabled preparation of this book. We also express our gratitude to Ms Maryam Jahani Kondori for helping us with the drawing figures given to this book and unifying the chapters.

Mr. Tajamul Showket's help in typesetting and preliminary editing is highly acknowledged.

We acknowledge the undying support we received from our families as that has been the fountainhead of inspiration and encouragement.

July 2006   **M. Kafi and Co-workers**

Contents

List of Contributors

Alizadeh A.
Faculty of Agriculture, Ferdowsi University of Mashhad, Iran

Bagheri A.
Faculty of Agriculture, Ferdowsi University of Mashhad, Iran

Hagian Shahri M.
Agriculture and Natural Research center of Khorasan, Iran

Hemmati Kakhki A.
Iranian Research Organization for Science & Technology, Khorasan Center, Mashhad, Iran

Kafi M.
Faculty of Agriculture, Ferdowsi University of Mashhad, Iran

Karbasi A.
Faculty of Agriculture, Zabol University, Iran

Mahmodi A.
Dryland Research Center of North Khorasan, Iran

Mollafilabi A.
Iranian Research Organization for Science & Technology, Khorasan Center, Mashhad, Iran

Rashed Mohassel M.H.
Faculty of Agriculture, Ferdowsi University of Mashhad, Iran

Sanuie Mohassel M.
Nader Production and Processing Company, Mashhad, Iran

CHAPTER 1

Historical Background, Regions of Production, and Applications of Cumin

M. Kafi
m.kafi@ferdowsi.um.ac.ir
Faculty of Agriculture, Ferdowsi University of Mashhad, Iran

1-1 HISTORICAL BACKGROUND

Cumin (*Cuminum cyminum*) is one of the most valuable medicinal herbs and spice in the world. This plant belongs to the *Apiaceae* family and is geographically distributed in south Mediterranean and West Asia (latitudes 20-38° N and longitudes 30°– 80° E) (6, 39, 42).

There are different theories about the origin of cumin. However, based on documented evidences, it either originated in northern Egypt, in the south Mediterranean climate or in the Middle East (10, 11, 28, 30). The presence of wild cumin plants in vast areas of south Mediterranean, Saudi Arabia, Iran, Central Asia, Sahara and south Pakistan indicate that these areas could be the origin of domestication of cumin (6, 28, 39). Based on the background of cultivation, variation of wild types, particularly other species of cumin (*Cuminum setifolium*) and wide distribution of semi-wild races of cumin, the Iranian plateau and Middle East could also be the center of evolution of cumin (20, 21, 28).

At present, cumin is cultivated in India, Iran, Argentina, Morocco, Ukraine, Egypt, Denmark, Lebanon, Malta, Mexico, Afghanistan, Pakistan, Turkey, Central America, Central Asia, China, Indonesia, with India and Iran being the main producers and exporters of this special spice worldwide (6, 15, 17, 39, 42).

The oldest evidence of utilization of cumin dates back to 5,000 years when it was used by the Egyptians as one of the main ingredients for mummification of the body of pharaohs (11, 15, 31, 42). Cumin is also symbolized as cupidity among Greeks, and persons who were miserly were jokingly said to have eaten cumin. Cumin is mentioned in Isaiah XXXVIII, 25 and 27.

The English word cumin is derived from the Latin word cuminum. There is no evidence of cumin before Islam came to Iran and the Middle East countries (611 AC). In the post-Islamic literature and in Iranian dictionaries and books, the Arabic word "Comon", Persian name "Zeera" or "Karavia" has been used both for the name of the plant as well as its product (12, 16, 20, 21). The common name of this plant is very similar in most countries, for instance the name of this plant in various languages is mentioned in Table 1.1 (3, 39).

Avicenna, the most famous traditional physician of Iran in 11th, century in his book Ghanoon (Laws) in Medicine, second volume, which is about medicinal plants and spices, wrote "There are many types of cumin some of which are Kermani, Farsi, black, yellow, Shami, and Nabti. Farsi cumin is stronger than Shami and Nabti. There are two types of cumin cultivation, wild type and domestic cumin. It grows in different types of soil" (3).

Khaje Nassiroddin Fazlolla, the minister of Mongol emperors in his book, *Alassar and Valahya*, in the 13th century also stated that Arabs call cumin, Komon, and grow it without irrigation. They believe that one should just give hope to the plant and anticipate that sometime the farm will be watered, however it does not need to be irrigated. Kerman province of Iran is the major producer of cumin, and it produces the best quality in the world because of its dry terrain, which must be irrigated regularly (11, 21).

Table 1.1   *Common names of cumin in different languages (39)*

Arabic	Kamoun, Kamun
Bengali	Jeera
Bulgarian	Kimion, Kimion italianski, Kimion rimski
Burmese	Ziya
Chinese (Mandarin)	Kuming, Xiao hui xiang, Zi ran
Croatian	Kumin
Danish	Spidskommen, Kloeftsvoeb
Dutch	Komijn, Djinten
English	Green cumin, White cumin, Cumin
Esperanto	Kumino
Farsi	Zireh, Zireh sabz
French	Cumin, Cumin blanc, Cumin du Maroc, Faux anis
German	Kreuzkümmel, Weißer Kreuzkümmel, Römischer Kümmel, Mutterkümmel
Greek	Kimino
Hindi	Jeera, Safaid jeera
Indonesian	Jinten, Jinten putih
Italian	Cumino, Cumino bianco
Japanese	Kumin, Umazeri
Portuguese	Cominho
Russian	Kmin, Kmin tminovyj, Kumin, Kyummel, Rimskij tmin, Zira, Kmin
Sanskrit	Jiira, Jiiraka, Jiirana, Sugandhan, Udgaarshodan
Spanish	Comino, Comino blanco
Turkish	Kimyon, Acem kimyonu, Kemnon
Urdu	Jirah, Zeera, Zira

Ayurveda, the science of life, which is the ancient traditional system of Indian medicine, is believed to be well over 5,000 years old. Ayurveda has particularly emphasized the subtle yet incredible healing properties of herbs and spices, and among these cumin plays an important part, which is used both as an ayurvedic medicine and as well as the main spice in Indian food recipes.

The state of Kerala in India is the major consumer of medicinal plants. The chief users of these are the Ayurvedic pharmacies mushrooming all over the state. The demand for ten major medicinal plants by the pharmacies is seen to be moving correspondingly with the rate of growth of consumer demand for the Ayurvedic medicines, which is 10% per annum from 1993 to 1996. The major ten plants are *Sida spp., Tinospora cordifolia, Asparagus racemosus, Phyllanthus emblica, Aegle marmelos, Terminalia chebula, Withania somnifera, Cuminum cyminum, Strobilanthes,* and *Adhatoda vasa* [A. *vasica*] (35).

1-2   USES OF CUMIN

Cumin seeds and its essential oil are important economic parts of the plant. In cumin producing countries, the seed is sold for indigenous consumption, but for exporting, to reduce volume and weight, essential oil is extracted from the seed and exported to the international market. While cumin is a well-known spice, it has many other uses in the food, pharmaceutical, cosmetic, perfumery and soap industries (9, 42).

1-2-1   Food Industry

Cumin has a long history of use as a spice and for its wonderful aroma and taste. During the recent years public interest towards using natural additives instead of synthetic chemicals has led to a breakthrough in using cumin as a natural flavoring agent in the food industry (1, 15). Today cumin is widely used in Indian, Mexican, Iranian and Middle Eastern meals, particularly in curries and rice preparations, cheese making, confectionery, alcoholic and non-alcoholic beverages. Cumin should be used in a specific quantity depending on the food type (37).

Cumin is one of the most typical spices of India, especially in the southern part. The seeds are used as a whole and are fried or dry roasted before usage. Adding cumin seeds in hot butterfat normally flavors legumes, especially lentils. Furthermore, the seeds form an inseparable part of curry powder. It is also an essential ingredient in the preparation of northern Indian *Tandoori* dishes. A survey in India revealed the average per capita consumption of spices including cumin to be 9.54 g/day and at that level, the nutrient contribution from spices ranged from 1.2 to 7.9% of an average adult Indian male's requirement for various nutrients (37).

Mineral compositions of cumin have been studied using inductively coupled plasma optical emission spectrometry (ICP-OES) after acid digestion. Fourteen elements were considered, which include heavy metals like iron (Fe), manganese (Mn), zinc (Zn), copper (Cu), cobalt (Co), nickel (Ni), molybdenum (Mo), lead (Pb), and chromium (Cr), alkaline earth metals like calcium (Ca) and magnesium (Mg), lighter element like aluminum (Al), and non-metals like silicon (Si) and phosphorus (P). A very strong linear correlation exists between Fe and Al contents in the spices. Zinc also correlates well with iron. This study provides a reliable account of the endogenic concentrations of a number of common elements including heavy metals present in cumin (18).

Its seeds and essential oils were used both as flavoring and a source of naturally occurring antioxidants in biscuits. Addition of Butylated Hydroxy Toluene (BHT, 0.0025 and 0.005%), or cumin essential oils (0.075, 0.15 and 0.225%), or whole ground cumin seeds (5, 10 and 15% and 1.9, 3.8 and 5.6%) to biscuits (on wheat flour weight basis) and storage at room temperature for eight months showed that there were variable decreases in the unsaturated fatty acid contents of the biscuits. The range of the decrease was 9.71-8.87% for biscuits mixed with BHT, and 8.41-7.81% for biscuits mixed with cumin essential oils, and 9.63-10.25% for biscuits mixed with cumin seeds. The stability of biscuits treated with cumin essential oils was extended to six months, on an average, over control. Whole ground cumin seeds did not have any antioxidant activities in biscuits (4).

1-2-2 Medicinal Uses

In traditional medications, cumin has several applications, such as for treatment of jaundice, dyspepsia and diarrhea, and has stomachic, diuretic, carminative, relaxant, digestion, stimulant, astringent and abortifacient properties (22, 31). Cumin seed and/or callus extracts and essential oils demonstrated antibacterial, antifungal, and antiviral. . The extracts inhibited bacteria (particularly *Staphylococcus aureus*) and fungi (*Fusarium moniliforme*) as well as polio and Coxsackie viruses. No significant anti-tumor activity was observed. The orally administered seed powder (2 g/kg) lowered the blood glucose levels in hyperglycemic rabbits (22).

Cumin has demonstrated antimicrobial effects; it retards the growth of mycelium and aflatoxin in *Aspergillus ochraceus*, *Coriolus versicolor*, and *Coriolus flavus*. It also has beneficial mutagenic and blood clotting effects (14, 26, 27).

The antimicrobial activities of the volatile oil of cumin and the active constituent, cuminaldehyde, were investigated by Shetty *et al.* (32), *Aspergillus niger*, *A. flavus*, *A. parasiticus*, *Penicillium chrysogenum*, *Saccharomyces cerevisiae* and *Candida utilis* were more sensitive to cumin volatile oil and cuminaldehyde than bacteria. Among gram-negative bacteria, *Escherichia coli* was the most sensitive to the volatile oil, while *Pseudomonas aeruginosa* was the most resistant.

Pharmacological properties of essential oils of nine species belonging to the Apiaceae family were studied in rats and mice, and LD50 values were determined in mice. Coriander, celery, parsley and cumin oils induced significant anti-inflammatory activities two, three and five hours after administration to rats. Cumin oil showed a marked antifungal activity against all selected strains of fungi (*Penicillium notatum*, *Aspergillus niger*, *A. fumigatus* and *Microsporum canis*) (1).

The antimicrobial activities of essential oils of cumin were investigated by the steam distillation. The findings showed that gram-negative bacteria were more sensitive to the spices than gram-positive bacteria. Cumin revealed antimicrobial effects against some gram-positive and gram-negative bacteria used in the study (26, 27).

The cumin essential oil was effective against *E.coli*, *Pseudomonas Aeruginosa* and *Salmonella* sp. and their inhibitory zones were 18, 10 and 23 mm, respectively (34).

The antifungal properties of extracts from cumin were evaluated against *Curvularia lunata, Fusarium oxysporum*, and *Alternaria alternata* by poisoned food technique. After 72 hours of incubation, the extracts significantly inhibited the growth of all fungi. The root extracts of *F. vulgare* and *Coriandrum sativum* were most toxic to *Curvularia lunata*. Flower and seed extracts from *Anethum graveolens* recorded the greatest inhibition of *Alternaria alternata*. All extracts showed relatively greater inhibitory effect on *F. oxysporum*, with stem extracts from *Anethum graveolens* exhibiting complete growth suppression (29).

The essential oils extracted from the seeds of seven spices including cumin have been studied for antibacterial activity against eight pathogenic bacteria (*Corynebacterium diphtheriae, Staphylococcus aureus, Streptococcus haemolyticus, Bacillus subtilis, Pseudomonas aeruginosa, Escherichia coli, Klebsiella spp.* and *Proteus vulgaris*), causing infections in the human body. It has been found that the oil of cumin is very effective against all tested bacteria. This oil is equally or more effective when compared with standard antibiotics, at a very low concentration (33).

The antibacterial activity against gram-positive (*Staphylococcus aureus, Bacillus subtilis* and *Streptococcus pyogenes*) and gram-negative (*Escherichia coli* pUC9 and *Pseudomonas aeruginosa*) bacteria, and antifungal activity against *Candida albicans* 174 of essential oils of cumin were investigated at 25 and 50 μg concentrations. Results were compared with some standard antibiotics (Imepenen, Amoxilin, Eritromycin and Ampicilin/Sulbactan). The gram-negative bacteria were more sensitive to the essential oils of both species than gram-positive bacteria. The *C. albicans* was also significantly inhibited by the essential oil treatment. Further results showed that the effect of standard antibiotics against the tested microorganisms were lower than the effects of cumin essential oils. This suggests that the essential oils of these species even at low concentration can be applied as anti-fungal and antibacterial agents (26).

Cumin is valuable in dyspepsia and diarrhea, and may relieve flatulence and colic. In the West it is now used mainly in veterinary medicine as a carminative, but it remains a traditional herbal remedy in the East. It is supposed to increase lactation and reduce nausea in pregnancy (5). Cumin is one of the main four ingredients of carminative syrup (Gripe Mixture), which is used extensively for children (31). Experiments using cumin on guinea chains and possible mechanism(s) proved the positive effect of cumin as a relaxant agent (8).

1-2-2-1   Application of Cumin Essential Oil as Antioxidant

Results of many research projects indicated that application of cumin essential oil prevents oxidation and decay of edible oils. The result of one such study proved that adding 2,000 ppm cumin and 1,000 ppm *Thymus vulgaris* essential oil to sunflower oil represents the same antioxidant effect as that of synthetic antioxidant.

Essential oil from cumin seed was fungistatic at low doses and fungicidal at high doses to aflatoxin-producing strains of *Aspergillus flavus* and *A. parasiticus* (10).

Cumin aldehyde of cumin essential oil contains a phenolic structure and active hydroxy group, which has antioxidant properties. Synthetic antioxidants have some injurious effects on the human body such as genetic mutations, toxicological and cancer side effects; therefore, efforts are underway to find a replacement for synthetic essential oils with natural essential oils, and cumin essential oil is one of these plant essential oils. Application of cumin and other plant essential oils in order to prevent oxidation and decay of edible oils and food materials containing oil has been the subject of many research projects (7, 9, 13, 19, 26, 27, 32).

Hemmati Kakhki and Mohazab Rahim Zadeh (19) applied essential oils extracted from cumin and *Thymus vulgaris* to liquid sunflower oil without antioxidants in three levels and studied the antioxidant effects of these essential oils with control, which was antioxidant free, and 200 ppm synthetic antioxidants, butylated hydroxytoluene (equal to antioxidants that is normally added to edible oils) after 75 days at 40°C. The result of this study showed that adding 2,000 ppm cumin and 1,000 *Thymus vulgaris* essential oil to sunflower oil represent the same antioxidant effect as that of synthetic antioxidant (Figs. 1-1 and 1-2).

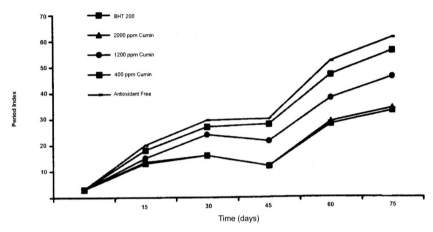

Fig. 1-1   Changes in peroxidase index in different treatments of sunflower oil with cumin and synthetic antioxidant application. The number beside each compound is the amount of application in ppm (19).

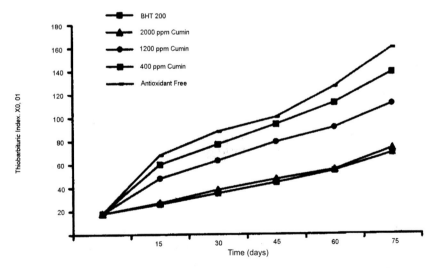

Fig. 1-2   Changes in thiobarbituric index in different treatments of sunflower oil with cumin and synthetic antioxidant application. The number beside each compound is the amount of application in ppm (19)

Essential oils of many other plants have antioxidant properties. Chipault (9) has indicated the antioxidant effects of essential oil in 22 different spices and aromatic plant species on lard fat at 98.6°C using active oxygen method. The results showed that *Thymus vulgaris*, *Rosmarinus officinalis* and *Salvia officinalis* expressed maximum antioxidant properties.

Wu and his colleagues (38) separated active materials from rosemary, oregano and cumin essential oil and introduced carnosic acid, oreganole and cumin aldehyde as the primary antioxidants compounds in these plants respectively. Farag and co-workers (13) studied the antioxidant strength of essential oils of six medicinal plants on oxidation of linoleic acid, application of synthetic antioxidant, butylated hydroxytoluene, and antioxidant free sample, and found that clove, rosemary and oregano are stronger, and salvia and caraway are weaker antioxidants than cumin.

Taha *et al.* (36) studied the effects of cumin and oregano essential oils as natural preservatives and compared them with oils which were – (a) preservative free, and (b) those with an application of synthetic preservatives, and measured the oxidation and lipolyze rate on different treatments. They reported that application of natural preservatives showed better results than other treatments.

Bassioiny (7) applied fresh powder and ether extraction of four medicinal plants including, cumin, marjoram, spearmint and basil as natural preservatives in a biscuit and found that adding ether extraction of these plants expressed stronger antioxidant properties than synthetic antioxidant, butylated hydroxytoluene.

Farag and co-workers (14) explored the antimicrobial properties of cumin and oregano on four gram-positive bacteria, three gram-negative bacteria, one acidophyle bacteria and one yeast and found that the sensitivity of gram-positive bacteria to these essential oils was higher than the rest. Therefore, cumin essential oil could be applied as an alternative antioxidant in place of synthetic antioxidants.

1-2-3   Use of Cumin Leaves as Animal Feed

One hectare of cumin field produces about 1.0 tons of shoot dry matter. Where cultivated areas are high (e.g. 200,000 ha in India and 47000 ha in Iran) considerable amount of shoot dry matter is produced annually from cumin fields, which could be used as dried forage. However, information on the food value of cumin leaves is scarce. Possibility of using this forage source for native animals (camel, sheep and goats) in Rajasthan state in India and Khorasan province in Iran should be studied (19). Traditionally farmers have been adding cumin straw to their animal nutrition and the result, particularly in lactating animals, is satisfactory. Low price of cumin forage and the possibility of enhancement of its quality make it a potential animal feed, however, further detailed studies on these aspects are required.

Feeding of the herbal galactagogue Payapro (comprising *Laptadenia* [Leptadenia] *reticulata, Nigella sativa, Foeniculum vulgare, Pueraria tuberosa, Glycerrhiza* [Glycyrrhiza] *glabra, Cuminum cyminum* and *Asparagus racemosus*) to healthy lactating buffaloes at 4 boluses/animal per day in feed for 16 days resulted in a significant (>5%) increase in milk yield compared with untreated animals over an observation period of 49 days. The effect of increased yield was evident within seven days of treatment, peaked at day 21 and was sustained until day 35. The treatment was found to be highly costeffective (23).

The effect of nine herbal feed additives on the health and productive performance of 20 male buffalo calves, supplemented with different herbal combinations containing *Cyperus scariosus, Cuminum cyminum* and *Foeniculum vulgare* in a 3:1:2 ratio (HC1, Group I); given at a rate of 10 g/100 kg live weight, was conducted. The N retained as percentage of absorbed N, i.e. apparent biological value was better in animals given herbal combinations (5).

The shoot of cumin contains essential oil. Essential oil of the seed and herb of cumin were extracted. In the herb and seed oils, 21 constituents were identified, representing 90.2 and 95.6% of the total amounts, respectively. Eleven components were similar in both herb and seed oils. Cumin aldehyde was found as the main component at concentrations of 53.6% for seed oil and 40.5% for herb oil. Qualitative and quantitative data indicated that oil production from cumin herb is a possibility (11).

1-3 ECONOMIC IMPORTANCE

In addition to food crops, man is meeting his many other requirements from plant products. Among these groups, paper plants, natural fibers, non-alcoholic drinks like tea, coffee and cocoa, and medicinal plants are of crucial importance (15, 41).

Cumin as one of the world's valuable spice/medicinal plant has great domestic consumption and is of immense exportable importance in Indian and Iranian agricultural economics (39, 40, 41). At present, India is the main producer of cumin followed by Iran. Table 1-2 shows the cultivated area, annual production and yield of cumin in the main cumin producing countries during the year 2001-2002 (24, 40, 41).

Table 1-2   *Cultivated area, production and yield of cumin in the main cumin producing countries during the year 2001-2002*

Country	Cultivated area (ha)	Annual production (tons)	Yield (kg ha^{-1})
India	381,534	145,110	262
Iran	50,000	16,250	325
Turkey	-	2,000	-
Syria	-	1,450	-

In cumin growing countries, such as India and Iran, the importance of cumin cultivation, is related to high water use efficiency, rural employment, and higher net profit compared to other crops, in addition to its value as an export commodity (2). Socio-economic studies have shown that on an average, production of 1 ha cumin in Iran (including all practices from sowing to harvest) is labor intensive and requires 35 persons/day. Given the total production area of 50,000 ha, it means 1,750,000 labor-days annually. Since 200 days employment is considered as a permanent job, cumin production in Iran provides up to 8,750 job opportunities per year and, therefore, is considered as an important factor against migration from rural areas (2, 10) and this figure in India with more than 300,000 ha under cumin cultivation, produces 52,500 job opportunities per year. .

1-4 ACREAGE, TOTAL PRODUCTION AND YIELD

At present, cumin production is limited to countries with ancient civilization and history, such as Egypt, Iran, India, Argentina, Morocco, Ukraine, Lebanon, Mexico, Afghanistan, Pakistan, Turkey, Central Asia, China and Indonesia. Data in Table 2-1 shows that India is the largest contributor, both in terms of cultivated area and quantity produced (40). Production trends in India and Iran are discussed in detail in the following paragraphs.

Figure 1-3 shows that total cultivated areas of cumin in Iran fluctuated from 50,000 to 11,000 ha during the last 10 years (24). A similar trend was also observed for total cumin production (Fig. 1-1). However, there is a negative relationship between the total area under cultivation and production, due to cultivation of some unfertile and marginal lands in the years where the amount of precipitation is higher than average in the cumin sowing season. The price of cumin in the previous year can also affect the area under cultivation in the successive year. Annual fluctuation in production and cultivated area (Fig. 1-1) compared with other winter crops like wheat and barley is much higher, which is mainly related to yield variation. It has been observed that after drought periods, farmers change the cropping pattern from cumin to some other crops like wheat and barley (24, 39).

Cumin is cultivated in arid and semi-arid areas without irrigation or in some areas with limited irrigation; therefore, the yield of this plant mainly depends upon the amount and distribution of rainfall. There are also some fungal diseases, such as *Fusarium oxysporum*, *Alternaria burnsii*, and *Erysiphe polygoni*, which in humid season, could affect cumin yield. Therefore, yield of cumin shows a remarkable fluctuation in different years, for instance during the last 20 years the yield of this plant varied from 104 to 412 kgha^{-1} in dry land farming and between 244 and 873 kgha^{-1} in irrigated farms in Iran (24). In India the same scenario was reported in cumin yield (39, 40, 41).

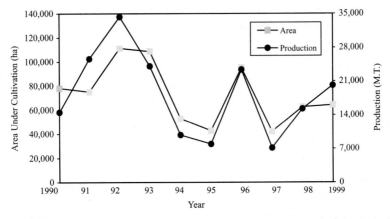

Fig. 1-3 Changes in area under cultivation and total production of cumin in Iran during the period 1990-1999

During the last few years, cumin production has spread to several other parts of Iran. Khorasan (mainly Central and South) province is known as the most important cumin production area of Iran (24, 25). Southern and Central parts of this province have unique climatic conditions for cumin growth and, therefore, 80% of the total country's production and 85% of cultivated areas of Iran belong to Khorasan. The major parts of Khorasan province where cumin is produced are Ferdows, Sabzevar, Gonabad, Taibad, Torbat-e-Jam, Kashmar and Birjand (13).

Statistics reveal that both the cultivated area of cumin and quantity produced in India (the main producer of cumin), are four times greater than that of Iran (the second largest producer). Rajasthan is the main state of cumin production in India, more than 90% of the area under cultivation and production of cumin is allocated in this state. Districts such as Barmer, Jalore, Nagaur, Pali,

Jodhpur, and Ajmer have the highest area under the cultivation of cumin. Gujarat is located adjacent to Rajasthan, but the total production of this state is less than 1,000 tons per year, while in Rajasthan more than 80,000 tons cumin seed is produced annually (41).

Figure 1-4 shows that total cultivated area of cumin in India ranged from 140,000 to 325,000 ha during years 2000-2004 (39). A similar trend was observed for total cumin production (Fig. 1-4) ranging between 30,000 and 140,000 MT during the same period.

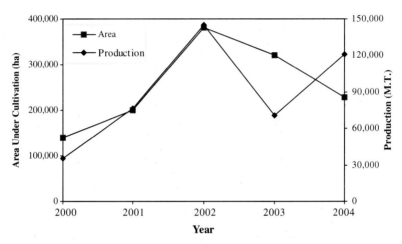

Fig. 1-4   Changes in area under cultivation and total production of cumin in India during the period 2000-2004

1-5 SUMMARY

Cumin is one of the most valuable medicinal and spice herbs in the world. Cultivation of cumin dates back to 5,000 years . This species is native to Egypt or south Mediterranean regions. The Iranian plateau and Middle East could also be the center of evolution of cumin. Cumin is cultivated in India, Iran, Argentina, Morocco, Ukraine, Egypt, Denmark, Lebanon, Malta, Mexico, Afghanistan, Pakistan, Turkey, Central America, Central Asia, China, Indonesia, etc. India is the main producer of cumin worldwide. The area under cultivation, as well as total production of cumin in India, is at least four times greater than that of Iran (second largest producer). Rajasthan is the main state of cumin production in India. In Iran, the main cumin production areas are located in Khorasan, a province in the south east of the country. Under optimum climatic conditions for growth and development of cumin and with its low water requirement, growth during winter and spring, and high net income gives cumin an advantage over many other crops. Cumin production is labor intensive and in highly populated rural areas provides permanent employment for many people. Cumin yield is sensitive to precipitation and fungal diseases, and shows high temporal variation in response to drought. While cumin is mainly known as a spice, it also has medicinal properties, such as for treatment of jaundice, dyspepsia, diarrhea, and has stomachic, diuretic, carminative, relaxant, digestive, stimulant, astringent and abortifacient properties, and antibacterial, antifungal, antiviral and anti-tumor activities. Antioxidant effects have been studied intensively during the recent years.

REFERENCES

1. Afifi, N.A., A. Ramadan, A. et al. 1994. Some pharmacological activities of essential oils of certain umbelliferous fruits. Veterinary Medical Journal Giza, 42(3): 85-92.

2. Atayee, M. 1977. Agronomy (Vol. 3). Tehran University Press. Tehran, Iran.

3. Avicenna (1049 A.D.) 1983. Ghanoon in Medicine. Vol. 2. (Translated by Sharaf Kandy A.) Shush Publishers. Tehran, Iran.

4. Badei, A.Z. and H.H. Hemeda, et al. 2000. Antioxidant activity of anise and cumin essential oils in stored biscuits. Egyptian Journal of Agricultural Research, 78(5): 2047-2065.

5. Bakshi, M.P.S. and M. Wadhwa. 2004. Effect of herbal feed additives on the nutrient utilization in buffalo calves. Bubalus Bubalis, 10(1): 65-70.

6. Balandary, A. 1992. Collection and survey of botanical characteristics of land races of Iranian cumin. Iranian Research Organisation for Science & Technology, Khorasan Centre.

7. Bassioiny, S.S. and F.R. Hassanien. 1990. Efficiency of antioxidants from natural sources in bakery products. Food Chem. 37: 297-305.

8. Boskabadi, M. H., S. Kiani and H. Azizi. 2005. Relaxant effect of *Cuminum cyminum* on guinea chains and possible mechanism(s). Indian Journal of Pharmacology. 37(2):111-115

9. Chipault, J.R. 1986. The antioxidant effects of spices in foods. Food Tech. 40: 209-211.

10. Dube S., P. D. Upadhyay and S.C. Tripathi 1991. Fungitoxic and insect repellent efficacy of some spices. Indian Phytopathology. 44 (1): 101-105.

11. El Sawi, S.A. and M.A. Mohamed. 2002. Cumin herbs as a new source of essential oils and its response to foliar spray with some micro elements. Food Chem. 77:75-80.

12. Estakhri, A. 1967. Almasalek and Almahalek (by the efforts of Iraj Afshar), Bank Melli Press. Tehran, Iran.

13. Farag, R.S., A.Z.M.A. Badei, R.M. Hewedi and G.S.A. El-Baroty. 1989. Antioxidant activity of some spice essential oils on linoleic acid oxidation in aqueous medi. J. Am. Oil Chem. Soc. 66: 792-799.

14. Farag, R.S., Z.Y. Daw, F.M. Hewedi and G.S.A. El-Baroty. 1989. Anti-microbial activity of some Egyptian spice essential oils. J. Food Protection. 52: 665-667.

15. Farrell. K.T. 1998. Spices condiments and seasoning. Culinary & Hospitality Industry Publications Services.

16. Fazllolah Hamadani, Kh. (1321 A.D.) 1989. Al Assar val Ahya (by the efforts of Setaude Afshar), Tehran University Press.

17. Guenter, E. 1982. The essential oils. Rebert E. Kpieger Publishing Company, Malabar, Florida.

18. Gupta K.K., S. Bhattacharjee, S. Kar, S. Chakrabarty, P. Thakur, G. Bhattacharyya and S. C. Srivastava 2003. Mineral compositions of eight common spices. Communications in Soil Science and Plant Analysis. 34(5-6): 681-693.

19. Hemmati Khakhki, A. and V.M. Mohazab Rahim Zadeh. 1999. Anti-oxidant effects of cumin and origanum essential oils on sunflower oil. Iranian Research Organisation for Science & Technology, Khorasan Center.

20. Heravi, A. (971 A.D.) 1967. Alabnia Anel haghayegh el Adviah (by the efforts of Bahmanyar A.). Tehran University press. Tehran, Iran.

21. Heravi Abu Nasr, Gh. (1542 A.D.) 1977. Ershado Zerah. Amir Kabir Publishers. Tehran, Iran.

22. Jain S.C., M. Purohit and R. Jain 1992. Pharmacological evaluation of *Cuminum cyminum*. Fitoterapia., 63(4): 291-294.

23. Khurana K.L., Balvinder Kumar, Sudhir Khanna, Anju Manuja, Kumar B., Khanna S. and Manuja A. 1996. Effect of herbal galactagogue Payapro on milk yield in lactating buffaloes. International Journal of Animal Sciences. 11: 1, 239-240.

24. Khorasan Agricultural Organization. 1991. Census of cumin production in different cities of Khorasan (in Persian). Mashhad, Iran.

25. Khorasan Standard and Industrial Research Organisation. 1976. 1987: Export of cumin from Khorasan. Mashhad, Iran

26. Kizil S. and Sogut T. 2003. Investigation of antibacterial effects of some spices. Crop Research Hisar. 25(1): 86-90.

27. Kizil S. and Sogut T. 2002. An investigation on antimicrobial effects of essential oils of coriander and cumin at different concentrations. Turkish Journal of Field Crops, 7(1): 1-5.

28. Mozafarian, V. 1983. Plant of Umbeliferae family. Ministry of Agriculture and Natural Resources. Tehran, Iran.

29. Prabha Purohit and Bohra A. 2002. Antifungal activity of various spice plants against phytopathogenic fungi. Advances in Plant Sciences, 15(2): 615-617.

30. Ray Chaudhari, S.P. 1992. Recent advances in medicinal aromatic and spice crops (Vol. 1). Today and Tomorrows Printer Publishers, New Delhi, India.

31. Sadeghi, B. 1991. Effects of different amount of nitrogen fertilizer and Irrigation in cumin production. Iranian Research Organisation for Science & Technology, Khorasan Center. Mashhad, Iran.

32. Shetty R.S., Singhal R.S. and Kulkarni P.R. 1994. Antimicrobial properties of cumin. World Journal of Microbiology-and-Biotechnology., 10: 2, 232-233.

33. Singh G. Kapoor I.P.S., Pandey S.K., Singh U.K. and Singh R.K. and 2002. Studies on essential oils: Part 10; Antibacterial activity of volatile oils of some spices. Phytotherapy Research, 16: 7, 680-682.

34. Stefanini M.B., Figueiredo R.O., Ming L.C., Junior A.F. and Szoke E. (ed.); Mathe I (ed.); Blunden G (ed.); Kery A. 2003. Antimicrobial activity of the essential oils of some spice herbs. Proceedings of the international conference on medicinal and aromatic plants, Budapest, Hungary, 8-11 July 2001. Part II. Acta Horticulturae. 2003, No.597, 215-216.

35. Suneetha M.S. and Chandrakanth M.G. 2002. Trade in medicinal plants in Kerala issues, problems and prospects. Journal of Medicinal and Aromatic Plant Sciences. 2002, 24: 3, 756-761.

36. Taha, S.H., R.S. Farag and M.V. Ali. 1990. Use of some essential oils as natural preservatives of butter. J. Am. Oil Chem. Soc. 68: 188-191.

37. Uma Pradeep K., Geervani P. and Eggum B.O. 1993. Common Indian spices: nutrient composition, consumption and contribution to dietary value. Plant Foods for Human Nutrition. 44: 2, 137-148.

38. Wu, J.W., M.H. Lee, Ho. Ct, S.S. Chang, 1982. Elucidation of the chemical structures of natural antioxidants isolated from rosemary. J. Am. Oil Chem. Soc. 59: 339-345.

39. www.uni-graz.at/˜katzer/engl/cumi_cym.html

40. www.indianspices.com

41. www.rajamb.com/cumin.htm

42. Zargari, A. 1988. Medicinal plants. University of Tehran Press. Tehran, Iran.

CHAPTER 2

Botany of Cumin Species

M.H. Rashed Mohassel
mhrashed@ferdowsi.um.ac.ir
Faculty of Agriculture, Ferdowsi University of Mashhad, Iran

2-1 INTRODUCTION

From the time of evolution till today, man has been using tools to cure his pain–the only difference is that in the primitive days he used simple crude tools, whereas nowadays with advanced technology, surgical equipment has become more sophisticated, but the basic objective remains the same (5). By introducing synthetic chemicals, the traditional medicine was almost forgotten, but during recent decades due to the occurrence of serious problems caused by these chemicals, use of medicinal plants has again been revived and is receiving special attention throughout the world.

Traditional medicine has historical roots in India and Iran. The age of Achaeinenian and Sasanian kingdom, especially after expansion of Islam right till the Mongolian governors, was the brilliant era for traditional medicine in both countries. Although Iranian doctors were the followers of Galen and Hippocrates, scientists such as Zakaria, Aliebne Abbase Majoosi and Avicenna (most famous doctors of their times), completed the work of Greek doctors and recorded valuable books in the history medicine. Some

European and American doctors during the 16th and 17th centuries and recently Arabists used to practice the Islamic traditional medicine, which opened up a new era in using medicinal herbs (5).

Using plants for medicinal purpose or as an additive or spice is still a common practice in small villages of Iran, India, Pakistan, and Arabian countries (8). The plant family Apiaceae is highly valued for its medicinal properties and uses as a spice. The seeds or pounded plant portions are used as additives in foods, or used in different forms as an antiflatulent, antigastritis, appetizer, etc. (5).

Cumin is an important medicinal herb in India and Iran and economically valuable as an export commodity, because it generates a remarkable amount of foreign exchange for the country (8). Description of the cumin family and the botany of cumin species are given below.

2-2 PARSLEY FAMILY *(APIACEAE)*

Annuals, biennials or perennials. Herbaceous, rarely undershrubs or shrubs with softwood and abnormal secondary growth. Different parts of plants aromatic, sometimes poisonous, mostly hollow stem or wide soft pith, and sometimes with several cavities in nodal areas (1). Parenchyma tissue with secretory ducts contains ethereal oils, resins and cumarins. Single secretory cells absent (7). In most plants of this family the stem is furrowed with dominant peripheral collenchymas (7, 19). Leaves alternate, rarely opposite or whorl, simple or compound as pinnate, palmate, ternate, and usually much divided, sheathing at the base and exstipulate.

Pedicels composed of a crescent or whorl of vascular bundles, sometimes with vascular bundles in pith area (7, 1). Inflorescence determinate as simple or compound umbels sometimes with cymes, raceme, panicle, or capitulum orientation, below each compound umbel a cluster of free or united bracts may be observed as involucres, below simple umbels a cluster of bracts may also be seen, called involucels (18). Flowers small, pentamerous, except gynoeciums, epigynous, bisexual, sometimes unisexual, rarely dioecious (1, 7).

Calyx five, free, regular, sometimes irregular or reduced as teeth above ovary, rarely absent (9). Corolla five, petals usually slightly folded inward, usually two-lobed, mainly white or yellow, sometimes purple or other colors. Aestivation imbricate or valvate. Stamens antipetal, located on nectary disc with filament (3); anther with four pollen sac, dehiscent with two sutures, protandrous. Pollination indirect. Gynoecia two locular (rarely unilocular) and two carpellate, ovary inferior placentation apical; stigma humid, capitate to elongate, two parted - sometimes up to five parts, style ± bulged at base to form stylopodium, connected to nectar disc (7); it seems that the origin of the stylopodium is mainly from the ovary rather than from the style (1). The lateral vascular bundles connected to ventral bundles of the ovary beneath the area that ovule connect to the ovary. Ovule anatropous (mostly one ovule, petals usually slightly folded inward aborted in each locule), ovary with a thin wall, sometimes may be thickened (9, 16).

Fruit schizocarp composed of two achenes' mericarp; mericarps are usually connected to a central intact or split carpophore, schizocarps are smooth or furrowed and sometimes covered with hairs, spines, etc. (4, 6, 7, 9, 10, 16, 18, 19). Mericarps are united when young but split apart when they dry (Fig. 2-1), each mericarp consists of five primary and secondary lobes. Embryo small, seeds endospermous. The testa is connected to the fruit, some of them may have drupaceous fruit. The general floral formula of this family is (3, 7):

$$\oplus \, \female \, K_5 \, C_5 \, A \, G_{\overline{(2)}} \text{ schizocarp or drupe}$$

Drupe fruits may disseminate by birds, but schizocarp fruits disseminate via wind or other means.

This family consists of 300 genera and 3,000 species. If Aralioideae species are considered within this family, there will be 460 genera and 4,250 species (7, 12, 13). The largest genera of this family are *Schefflera* (600 species), *Eryngium* (230 species), *Polycias* (200 species), *Ferula* (150 species), *Pecedanum* (150 species), *Pimpinella* (150 species), and *Bupleurum* (100 species) (7).

This family may be divided into the following subfamilies:

Apioideae: Consists of most common umbelliferous plants in Iran and the northern temperate hemisphere. These are herbaceous plants with compound umbel inflorescence and schizocarp fruit. The floral formula of this subfamily is:

$$\oplus \, \male\female \, C5 \, K_5 \, A_5 \, G_{\overline{(2)}} \; \text{schizocarp}$$

The *Apiaceae* usually refer to this subfamily. The species within this group have secretory canals with essential oils. Figure 2-1 illustrates the floral diagram and different parts of cumin stem and fruit as an example of *Apioideae*.

Saniculoideae: Consist of stylopodium separated from the style via a narrow groove (e.g. *Eryngium*). These two groups are closely related because they are herbaceous, having essential oils in fruit and the order of mat k (maturase controller gene in chloroplast), and *rbcL* (the gene which controls most functions of rubisco, the carbon acceptor in photosynthetic eukaryotes and cyanobacteria).

Aralioideae: Sometimes taxonomists place this group within the family called *Araliaceae*.

Presently, the monophyletic nature of *Apiaceae* is documentary due to morphological characteristics, secondary metabolites and the order of rbcL and mat k (2, 7). It is closely related to *Pittosporaceae*, and these families along with two or three other small families are within *Apiales*. The presence of resin, essential oils, within secretory ducts of conductive tissues, the orientation of adventitious roots, and the presence of falcarinons polyacetylenes, small embryo and bractaceuse type leaves at the base of the stem are characteristics that differentiate these two families from each other. They constitute the central core of Astrid because they have an ovule with a single integument, usually a thin layer of megasporangium, and united corolla (pronounced in *Pittosporaceae*). The stamens are located on a ring, the order of *rbcL* and *Atp B* (a gene which controls β subunit in ATP syntheses), and the presence of DNA within chloroplast. All these groups definitely possess oxyacetylene compounds (12, 13).

Traditionally and concisely, true *Apiaceae* is believed to consist of an elongated carpel at each side of the ovary, schizocarp fruits with carpophores, and having essential oils and secretory ducts. It also consists of woody species with two to five carpels and drupaceous fruit without active oil, and belongs to *Aralioidae*, also known as *Araliaceae* (13, 17).

Hydrodicotyloideae: With lignified endocarp and without free carpophores (1), may be recognized as an intermediate between Araliaceae and Apiaceae (12, 13). Some people also believe that either *araliaceae* or *Hydrodicotyloideae* may be studied in separate families.

These plants are mostly distributed in the Mediterranean region in Iran, Turkey and central Asia, where *Apiaceous* plants are considered to have originated.

The species of this family are highly valuable from many aspects, particularly medicinal and nutritional usage including:

Dill *(Anethum graveolens)*, celery *(Apium graveolens)*, carrot *(Daucus carrota)*, *Falcaria vulgaris*, *Heracleum Persicum*, coriander *(Corianderum sativum)*, parsley *(peteroselinum crispum)*, black cumin *(Bunium persicum)*, caraway *(carum carvi)*, cumin *(cuminum cyminum)*, fennel *(Foeniculum vulgar)*, and *Trachyospermum copticum*.

Most of the above mentioned plants are widely grown in parts of Iran and used for their medicinal or nutritional purposes (16). Iran is one of the main producers of cumin *(Cuminum cyminum)*. Cumin is cultivated in Iran, and Khorasan has the highest acreage cumin among different states of Iran. Unfortunately, despite being an ancient crop cumin could not find its real position between crops and is among forgotten plants. A thorough investigation of this crop from different aspects is necessary for us to review this forgotten plant. Therefore, in the following sections, the botany of the genus and species grown in Iran is reviewed.

2-3 CUMIN

Annual, aromatic, pubescent (except fruit). Stem herbaceous with dichotomous or sometimes trichotomous branches, 10 to 50 cm tall.

Stem furrowed with dominant peripheral collenchyma, the vascular bundles contain tracheid elements with simple pits. Parenchyma tissues have ducts secreting various substances especially ethereal oils, resins, terpenoids, saponins and monoterpens (1). Solitary secretory cells are not present in this genus. Leaves alternate, biternate and filiform (11, 4). Petiole with broad sheath and a ring of vascular bundles at the base, stipule absent (16).

Inflorescence a compound umbel composed of three to six radially single umbel each of which has three to five flowers. Bracts linear and elongated around the base of the compound umbel called involucres; the bract of each flower also arranged around single umbels called involucels. Flowers bisexual and complete five merous (except two carpelate ovaries). The flower formula is as follows:

$$\oplus \, \female \; K_5 \; C_5 \; A_5 \; G_{\overline{(2)}} \; \text{schizocarp}$$

Sepals and petals five, united to ovary, the teeth of sepals are pronounced, hooked and persistent. Perianth ± unequal. Corolla red, or whitish, tips slightly folded inward. Aestivation valvate (9, 16). Stamens five, bent and dehisce inward (9, 16), anti patulous, located on rector, ferrous disc via filament. Gynoecium two carpelate ovary inferior, two united locular, fruit fusiform and schizocarp composed of mericarp. Seed embryo small, enclosed within endosperm and tightly connected to mericarp wall, ventral portion of mericarp flat and dorsal portion concave, five longitudinal linear ridges on each mericarp, one dorsal, two lateral, and two marginal, each ridge with one vascular bundle, xylem inside and phloem outside the vascular bundle. Fruit 2mm x 5mm and covered with hooked hairs which differ from specie to specie (4), pericarp enriched with tannins and stains caused by ferrous salts. Fruit color yellow to grayish yellow. Figure 2-1 shows flower diagram, cumin fruit and cross-section of ovary in cumin. This genus in Iran includes two species which can be distinguished from each other by using the following key.

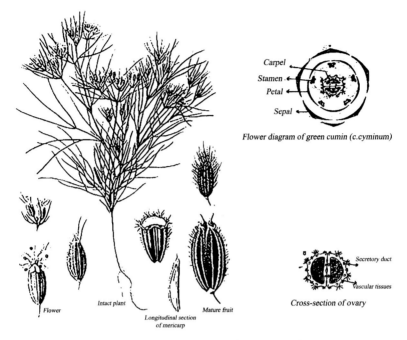

Flower diagram of green cumin (c.cyminum)

Carpel
Stamen
Petal
Sepal

Secretory duct
Vascular tissues

Cross-section of ovary

Flower

Intact plant

Mature fruit

Longitudinal section
of mericarp

Fig. 2-1   Different parts of green cumin

2-3-1 Key to Cumin Species

1- Corolla white, Fruit with long hairs .... *C. set folium*

2- Corolla red, Fruit with no hairs or small hairs .... *C. cyminum*

Sometimes *C. setifolium* is considered a subspecies of *C. cyminum* (16).

2-4 WHITE CUMIN
(*CUMINUM SETIFOLIUM* BOISS, KOS, POL)

Syn. *Psammogeton setifolia* Boiss

Torilis setifolia Boiss.

Annual, herbaceous, 10 to 30 cm tall with branched stem. Leaves with membranous sheath at base, blade filiform and ternate twice (1). Umbels radial, 3 to 4.5 cm long. Involucres and involucels filiform and branched. Fruit schizocarp composed of 2 achene mericarps, 4 to

5 mm long, covered with long dense hairs, 5 to 7 mm long. A picture of white cumin plant and a close-up of fruits are shown in Fig. 2-2.

This species grows in the wild in different parts of Iran including: Adzedshahr (864 m alt.) 32 km to Borazjasn (905 m alt.), Khorasan: in khowr to posht badam region (1,000 m alt.), Tabas (573 m alt.), Kavir area of Azbago (265m alt.), Damghan to Sabzevar (2,518 m alt.), Quchan Alam Ali high hills. (1,600 m alt.) and in Mardabad south of Karaj (1,300 m alt.) (14, 15,16).

The habitat of C. *setifolium* is similar to C. *cyminum* and at a glance they may look identical. The basic differences between them are the size of hairs which are longer in C. *setfolium,* and persistent calyx observed in these plants and *Psammogeton.* However, in *Psammogeton* the fruit hairs are glandular. C. *setifolium* also bears a similarity to *Ammodaucus leuctrichus* cosson, and Dur or *Cuminum maroccanum* Davis and Hedge, especially its subspecies *Nanocarpus beltian* growing in South Africa.

2-5   GREEN CUMIN *(CUMINUM CYMINUM* L.*)* SYN. *C. OFFICINALE* GARSAULT, DESCR. *C. ODOROUM* SALISB

C. *hispanicum* Bunge

Ligusticum Cuminum Grams

Cuminia cyminum J.F. Gmelin

Luersswnia cyminum D.Kuntze

Annual, herbaceous with very fine foliage. 15 to 50 cm tall depending on environmental conditions. Long, white and slender tap root. Stem herbaceous, dichotomously or rarely trichotomously branched, terminated into compound umbellous inflorescence. Leaves alternate, dark green, bright and glabrous, filiform and biternate (11), petiole connected to stem with pronounced sheath. Involucres around the main umbel consist of five to six thread type bracts, 5-10 mm long; involucels below and around each single umbel consisting of linear bracteoles (16). Figure 2-2 shows different stages

Fig. 2-2 *Cuminum setifolium,* A= intact plant, B= a cluster of fruits with short hairs

Fig 2-3 *Cyminum cuminun,* A= intact plant, B= a cluster of fruits

of cumin growth. Sepals green, five, free, sharp and pointed. Petals red to pink, ±unequal, and free. Stamens five. Ovary with two united carpels. Fruit schizocarp with two mericarps, mericarps united first but split apart when they dehydrate and mature.

The fruit is about 5 to 6 mm long, 1.5 – 2 mm wide, elongated and fusiform. Two parts of schizocarp are connected with a peduncle, the peduncle is split apart towards the ventral side of mericarp. Five prominent longitudinal ridges observed on each mericarp (4). Fruit extremely fragrant, covered with rough hairs, but sometimes without hairs. Pericarp has an enormous amount of tannins and stains with ferrous compounds, the color is yellow to yellowish brown or gray; secretory ducts are obvious in cross-section and longitudinal section of fruit. A picture of green cumin and a close-up of fruits are shown in Fig. 2-3.

Pollination in this plant is by wind and rarely insects. Umbels produce seeds from the bottom to top and from the external part of umbels towards the center, which indicates indeterminate inflorescence. Perhaps, under proper conditions the cumin is able to repeat production of flowers and seed.

Fruit of cumin is the important part of the plant with high usage value. It includes oil (7%), resin (13%), essential oil (2.5 to 4%), and aleuron. Essential oil is produced by distillation of mashed fruit, and has a very strong aroma with a mass volume of 0.91 to 0.93 (19). It dissolves in ten-fold of its volume at 80 °C and stronger alcohol. The essential oil is composed of cuminaldehyde or cuminol ($C_{10}H_{12}O$). The characteristics aroma of the essential oil and fruit is due to cuminol. In cumin essential oil, other chemicals such as cymene phllandrene, carvone, cuminiqué alcohol, etc. are present in small quantities (7). The storage and maintenance of essential oils should be in completely filled bottles and kept away from light and aeration. Cumin pulp is also rich in vitamins and consists of 18.7% protein, 86% carbohydrate, 10% oil plus minerals such as Ca, P, and Fe salts (1).

2-6 SUMMARY

Cumin, a plant from *Apiaceae*, is one of the most advanced species among Angiosperms. It is extremely valuable for its medicinal and nutritional properties, and is a source of income for farmers in the dry regions of Iran, particularly in parts of Khorasan . Exporting the grain of this crop generates a considerable amount of foreign exchange for the country. In Iran two species of cumin are grown, *C. setifolium* and *C. cyminum,* out of which the former is a volunteer plant in cumin plantations or cultivated areas, while the latter is mostly cultivated. The main difference between these two species is the presence of hair on the grain which is longer in *C. setifolium.* Considering the importance of cumin as a special crop, the botany of cumin was reviewed and investigated.

REFERENCES

1. Cronquist, A. 1981. An integrated system of classification of flowering plants. Columbia University Press. New York. Pp. 846-849.

2. Downie, S. R., S. Ramanth, D. S. Katz-Downie and E. Lianas. 1998. Molecular systematics of *Apiaceae* subfamily *Apioideae:* Phylogenetic analysis of nuclear ribosomal DNA internal transcribed spacer and plastid RPOCI intron sequences. Amer. J. Bot. 85:563-591.

3. Duta, A. C. 1979. Botany, for degree students. Oxford University Press. Calcutta, India. Pp. 735- 737.

4. Ghahraman, A. 1993. Iran cormophytes. Volume 2. University Center Publication, Tehran, Iran.

5. Hosseini- Tabib, M. M. 1999. Tohfeh Hakim momen. Sandoogh Press, Tehran, Iran.

6. Jackson, G. A. 1933. A study of carpophore of the *umbelliferae.* Amer. J. Bot.20: 121- 144.

7. Judd, W. S., C. S. Campbell, E. A. Kellogg and P. F. Stevens. 1999. Plant systematic. A phylogenic approach. Sinauer Associates Inc. Sunderland, Massachusetts USA. pp. 378- 390.

8. Kafi, M. and M. H. Rashed. 1990. The effect of weeding times and population density on yield and yield component of cumin. Journal of Science and Technology. Volume 6(2). 151- 158.

9. Mozaffarian, V. 2000. Plant classification, Book 2, *Dicotyledons*, Amir Kabir Publication Institute, Tehran, Iran.

10. Nath, R. 1984. Plant taxonomy, principles, advances, representative families and plants. Pub. by Metropolitan Book Co Ltd. Pvt, New Delhi, India. Pp. 295- 299.

11. Omid Baigi, R. 1998. Production and processing of medicinal plants. Volume 3, Astane Ghods Publication, Mashhad, Iran.

12. Plunkett, G. M., D. E. Soltis and P. E. Soltis. 1996. Higher level relationship of *Apiales* (*Apiaceae* and *Araliaceae*) based on phylogenetic analysis of *rbcl* sequences. Amer. J. Bot. 83: 499-515.

13. Plunkett, G.M., D.E. Soltis, and P. E. Soltis. 1997. Clarification of the relationships between *Apiaceae* and *Araliaceae* based on mat and rbcl sequences. Amer. J. Bot. 84: 567- 580.

14. Rashed-Mohassel, M.H. 1992. Flora of Khorasan. 1st report, Ferdowsi University Press, Mashhad, Iran.

15. Rashed-Mohassel, M.H. 1992. Flora of Khorasan. 2nd report, Ferdowsi University Press, Mashhad, Iran.

16. *Rechinger, K. 1981. Flora Iranica* . Apiaceae. Vol.162. Academische Druck-u-verganstalt, Graz, Austria. Pp. 140-142.

17. *Thorne, R.F. 1973. Inclusion of the Apiaceae* (Umbelliferae) *in the* Araliaceae. Notes Roy. Bot. Gard, Edinburgh. 32: 161-165.

18. Zargari, A. *1962.* The method of Plant Identification. Apetals and polypetals. Amir Kabir Publication Institute, Tehran, Iran.

19. Zargari, A. 1993. Medicinal Plants. Volume 2, Tehran University Press, Tehran, Iran.

CHAPTER 3

Ecophysiology of Cumin

M. Kafi
m.kafi@ferdowsi.um.ac.ir
Faculty of Agriculture, Ferdowsi University of Mashhad, Iran

3-1 GROWTH ENVIRONMENT

Cumin has been regarded as a temperate plant (12, 23). This species is successfully grown in the Middle East, North Africa and in parts of Asia and Europe, and its growing period completes before severe hot temperatures of late spring and early summer (8, 16, 18, 26). Optimum growth temperature of cumin ranges between 9° and 26°C, which usually coincides with late winter and early spring in many cumin growing areas. Therefore, in cumin producing countries such as Iran and India, this species is grown as a winter crop (22). Favorable environmental conditions for cumin production prevail in many parts of Iran. However, high relative humidity in the Northern parts of the country is the main growth limiting factor due to sensitivity of cumin to fungal pathogens such as *Fusarium oxysporum*, *F. cerumini*, *Alternaria burnsii* and *Erysiphe polygoni* which can easily spread at high levels of relative humidity (7, 13, 15). Therefore, cumin growing areas of Iran are restricted to dry regions adjacent to Central Desert (e.g. Khorasan, Isfahan and Kerman provinces). In India also relatively dry state, Rajasthan is the main state of cumin production, more than 90% of the area under cultivation and production of cumin is allocated in this state. Districts such as Barmer, Jalore, Nagaur, Pali, Jodhpur, and Ajmer have the highest area under the cultivation of cumin.

3-2 TEMPERATURE REQUIREMENTS

Cumin is known as a tropic or semi-tropic species by many researchers (22). However, the natural habitat of this species is not located in the tropics. The most probable centers of origin of cumin are Egypt, Turkmenistan and East Mediterranean (22). Therefore, the Mediterranean climate is most suitable for its optimal growth and development. Information about the base temperature of cumin is not fully documented but the scattered existing reports show that cumin seed does not germinate at temperatures close to 0 °C. Raychaudhuri (22) reported 9 °C and 26°C respectively, as minimum and maximum temperatures for cumin growth. The minimum temperature for seedling emergence is reported to be 2-5°C (24). While there is no report on the maximum temperature that cumin can withstand, it seems that this species cannot tolerate even the moderate summers of Mediterranean climates. High temperatures usually have no direct effects on cumin but may reduce the growth period and lead to early ripening. In such conditions the number of umbrellas per plant, seeds per umbrella and seed weight are reduced considerably. In low temperatures, cumin leaves change from green to purple and die due to long exposure to cold weather. In standard germination tests, a temperature range of 20-30°C is proposed for cumin (1). At a constant temperature of 17°C under controlled conditions, 100% germination was reported after 14 days (Kafi, unpublished data).

3-3 WATER REQUIREMENTS

Cumin is highly sensitive to *Alternaria* blight and *Fusarium* wilt and severity of the disease is closely related to moisture conditions (13, 15,

24). In wet years with high spring rainfall, damage due to this fungal disease is extremely high and crop yield may be reduced considerably. Therefore, water requirements of cumin should be scheduled in relation to these pathogens.

The natural habitat of this species in Iran, India and other production areas of the world is characterized by low relative humidity. Mean relative humidity of spring in some major cumin growing areas of Khorasan province is shown in Table 1-3.

Reports on optimum rainfall for cumin growth are highly variable. Available literature (3, 21, 24) reported an annual rainfall as high as 800-2,700 mm. Sadeghi (24) in a three-year experiment on the effects of irrigation and nitrogen fertilizers on cumin yield showed that while spring rainfall in three successive years was 26.1, 43.2 and 113 mm, irrigation treatments had no significant effect on yield, indicating low water requirement of cumin. In the last year of this study when spring precipitation was high, extensive spread of *Alternaria* blight was observed. (Details on irrigation requirements of cumin are discussed in Chapter 5.)

Table 3-1   *Mean relative humidity over a 10-year period for different weather stations during the month of maximum growth rate of cumin*

Station/Month	April†			May			June		
	MNG*	NON*	AFN*	MNG	NON	AFN	MNG	NON	AFN
Sabzevar	54.1	31.2	34.6	45.3	26.2	27.5	36.9	21.2	19.9
Mashhad	81.9	43.9	57.8	73.3	39.1	48.3	59.3	26.7	31.2
Birjand	52.0	21.0	28.0	40.4	20.3	20.4	28.6	13.7	15.9

*) MNG = Morning, NON = Noon, AFN = Afternoon
†) Data source: Meteorological Yearbook, Iran Meteorological Organization

3-4 SOIL

Small seedlings of cumin have low vigor and their emergence from heavy soils, especially when crusting occurs at the soil surface is difficult, therefore, sandy loam soils are recommended for proper stable establishment (7). Sadeghi (24) reported good establishment of seedlings in medium textured soils.

Sufficient aeration and good oxygen availability are crucial for successful growth of cumin (7). It is stated that cumin growth is optimal in soils with pH between 4.5 and 8.3, proper drainage, and sandy-loam texture (22).

Since biomass of cumin is too low compared to many other crops, its nutritional requirements are also low. For example, wheat crop with a grain yield of 5,000 kg ha^{-1} and harvest index of 0.4 produces 12.5 tons ha^{-1} dry matter, while, cumin with 500 kg ha^{-1} grain yield and a harvest index of 0.5 produces only 1,000 kg ha^{-1} dry matter.

Experimental results on soils with 0.44 –0.58% organic matter, 30-36 ppm P, 175-335 ppm K and pH of 7.8-8.2 showed that application of N fertilizers during three successive years had no significant effects on cumin yield (8). Fotovat (10) reported that cumin seed yield was not significantly influenced by application of N, P and K fertilizers up to 50, 20 and 25 kg ha^{-1} respectively, compared with control. Based on these results, he concluded that there is no need for fertilizer application in cumin fields if there has been fertilizer application for the previous crop. On the other hand, Patel et al. (19) reported 90 kg ha^{-1} increase in cumin yield with application of 40 kg ha^{-1} N and P$_2$O$_5$ fertilizers. However, water use efficiency was reduced with increasing fertilizer levels. In the work of Champawat and Pathak (6), application of N, P and K fertilizers at rates of 30, 20, and 30 kg ha^{-1} respectively, were the best fertilizer levels for maximum cumin yield and minimum prevalence of damping-off disease.

Sadeghi (24) reported a positive response of cumin to calcium compounds when fields were irrigated with wastewater of sugar beet manufactories containing up to 0.25% calcium carbonate.

3-5 LIGHT REQUIREMENTS

Leaf area index (LAI) of cumin is low and rarely exceeds 1.5 (15). Therefore, its open canopy absorbs a low portion of the incoming solar radiation. As a result, weeds can easily dominate in cumin fields due to their higher light competition ability (15). Information on optimum intensity and quality of light for cumin growth is scarce.

This necessitates further investigations on canopy structure and optimum LAI of plant under different radiation regimes.

3-6   YIELD COMPONENTS

Yield components of cumin consist of the number of plants per unit area, number of seeds per plant (the products of number of umbrellas per plant and number of seeds per umbrella) and seed weight. Due to sensitivity of cumin to seeding depth, planting date and physical and chemical conditions of soil, establishment of optimum crop density is the most important yield-determining factor. However, larger number of umbrellas per plant or larger number of seeds per umbrella may compensate for low crop density, but this may not fully compensate yield reduction due to low crop densities. Optimum plant density for cumin is about 120 plants m^{-2} (16), which could be obtained in mechanized cropping systems by adjustment of row distance at 40 cm.

Number of umbrellas per plant is the second important yield component of cumin. Aminpour and Mosavi (3) showed that the number of umbrellas per plant alone accounted for about 96% of variation in cumin grain yield, and grain yield (GY, kg ha^{-1}) could be estimated from the number of umbrellas per plant (NU) using a linear model:

$$GY = 496.28 + 23.49\,NU$$

In this experiment, the number of umbrellas $plant^{-1}$ varied from 19.9 to 55.1 depending on irrigation intervals.

Kafi and Rashed (16) showed that in a wide range of plant densities, the number of umbrellas per plant changed from 18.9 to 31.3 (Table 2-3). Salehi et al. (23) reported 27 umbrellas $plant^{-1}$ as an average value for cumin. Ehtramian (8) reported that the number of umbrellas per plant varies from 17.1 to 42.2 in a range of planting dates with a correlation coefficient of 0.22 between seed yield and number of umbrellas per plant. Nabizadeh (18) reported that the number of umbrellas per plant was reduced from 15 to 1 with increasing salinity levels (Fig. 3-1). Jangir and Singii (14) stated that under different nitrogen fertilizers and irrigation regimes, the number of umbrellas per plant is highly variable, ranging from 16.6 to 35.8. The number of cumin seeds per umbrella is a function of plant density and number of umbrellas per plant, and is highly dependent on environmental conditions at anthesis and early stages of seed formation. However, the importance of number of seeds per plant as a yield component of cumin is much lower than the number of umbrellas per plant.

Table 3-2 *Yield components of cumin in different crop densities*

Density (plant m^{-2})	No. of umbrellas plant^{-1}	No. of seeds plant^{-1}	1,000 grain weight (g)
40	31.3	16.8	2.87
80	23.4	12.4	2.99
120	18.9	11.3	2.79
LSD 0.05	4.8	1.9	0.66
Mean	24.5	13.5	2.88

Generally, the number of seeds per umbrella has an average value of 14 and usually does not exceed 28 (Table 3-2), but a wide range of number of seeds per plant has been reported by different researchers (14, 16, 23, 24). From the above results it could be concluded that number of seeds per umbrella is highly variable under different management practices such as planting date, irrigation intervals, nitrogen application, weeding intervals, salinity and water stresses (Table 3-3).

Table 3-3 *Variation of number of seeds umbrella^{-1} and 1,000-seed weight reported for cumin under different experimental conditions*

Treatments	1,000-seed weight (g)	Number of seeds per umbrella	Experimental environment	Reference
Without weeding	2.63	9.6	Field	13
Weeding (once)	2.99	14.4	Field	13
No salinity control	2.80	17.0	Greenhouse	17
Salinity (200 mol m^{-3})	0.80	1.0	Greenhouse	17
Nitrogen (50 kg ha^{-1})	4.03	11.45	Field	10
Nitrogen (200 kg ha^{-1})	4.01	12.48	Field	10
Irrigation (once)	3.24	30.3	Field	2
Irrigation (4 times)	3.04	22.6	Field	2
Mean	2.94	14.9	-	-

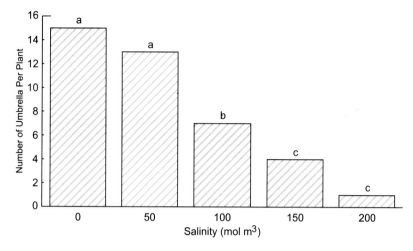

Fig. 3-1 Number of umbrellas per plant in cumin as affected by different salinity levels. a, b, c, d : Means with similar letters are not significantly different at 5% probability level.

Cumin seed weight is controlled mainly by genetic criterion. Therefore, seed weight is hardly influenced by environmental stresses or agronomic practices. In fact, cumin adjusts seed weight mainly by reducing seeds per plant, however, variation of cumin seed weight under extreme environmental conditions is evident (Table 3-3).

In the experiment of Aminpour and Mosavi (3) 1,000-seed weight at four irrigation treatments varied between 3.04 and 3.24 g (Table 3-3). Under different nitrogen applications and planting dates, 1,000-seed weight of cumin was between 3.6 to 5.1 g and a correlation coefficient of 0.99 was reported between seed weight and yield (8). Kafi (15) in an experiment with crop density and weed control intervals showed that 1,000-seed weight of cumin was relatively constant and changed form 2.79 to 2.99 g, whereas in the experiment of Jangir and Singii (14) cumin 1,000-seed weight varied between 4.03 to 5.20 g.

3-7 BIOLOGICAL YIELD

Biological yield is total harvestable dry matter produced by plants during the growing season, excluding roots. Therefore, it can be calculated as the product of mean crop growth rate (CGR) and length of growth period. While growth duration of cumin (planting to harvest) may last up to 150 days (8, 23), a large portion of this period (40-50 days) is for seedling emergence, thus the actual period of emergence to maturity is about 80-100 days (15, 23). On this basis, with 100 days of growth period and mean CGR of 3 g m^{-2} day^{-1}, biological yield of cumin is about 3 tons ha^{-1}. Length of growth period, mean CGR, and biological yield of cumin under different experimental conditions are given in Table 3-4.

Biological yield of cumin can also be estimated from CGR components, i.e. leaf area index (LAI) and net assimilation rate (NAR). Experimental results have shown that average LAI of cumin is 1 or lower (8, 16, 18, 23) with a mean NAR of 3-4 g m^{-2} leaf day^{-1} (15, 23). These evidences indicate that cumin has low potential for biomass production compared with many other field crops. Therefore, evaluation of possibilities for increasing cumin biomass by establishing higher LAI, increased leaf area duration and improvement of NAR are the main research priorities in cumin production. For example, rapid canopy closure with optimum LAI would be possible using practices such as increasing crop density or reducing between-row distance. Photographs 3-1 and 3-2 (Appendix) show cumin fields sown at 20 and 30 cm inter-row intervals with reasonable growth. However, it is obvious from photographs that even at the final stages of growth, canopies are rather open.

Table 3-4 *Length of growth period (emergence to maturity), crop growth rate (CGR, g m^{-2} day^{-1}), biological yield (kg ha^{-1}) and grain yield (kg ha^{-1}) of cumin reported by different researchers*

	Length of growing season (days)	Emergence to maturity (days)	CGR	Biological yield	Grain yield	Ref.
	100	85	3	1,300	666	13
	120	100	-	1,230	517	1
	100	80	-	1,750	770	17
	170	80	2.7	1,606	945	7
Mean	123	86	-	1,471	725	-

3-8 GRAIN YIELD

Cumin grain yield could be estimated from its components. Considering the highest values reported for each component and multiplying them, the maximum grain yield of 2,000 kg ha^{-1} will be obtained which is much lower than many crop species. The main reason for low yield potential of cumin is short growing period as well as its slow growth at seedling stage (Table 3-4). Maximum cumin grain yield of 1,800 kg ha^{-1} under experimental conditions with four times irrigation was reported in Isfahan, Iran (3). Other reported yields are usually lower than 1,000 kg ha^{-1} (8, 16, 23, 24). Considering that during more that 70% of growth period LAI of cumin is lower than 1 with a CGR below 1 g m^{-2} day^{-1}, this low grain yield is not surprising (Fig. 3-2).

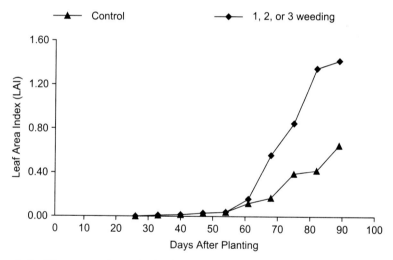

Fig. 3-2 Changes of cumin leaf area index under different weed control treatments. Since there was no significant difference between LAI in 1, 2, and 3 times weeding, these treatments were pooled.

Theoretically, if enough biomass is produced during vegetative growth period, grain yield increase would be possible by increasing seed growth rate or duration of grain filling period.

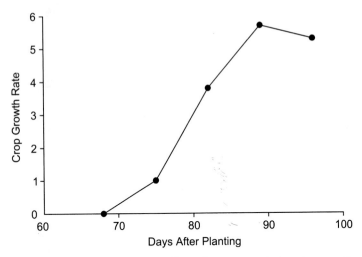

Fig. 3-3 Crop growth rate (CGR, g m^{-2} day^{-1}) of cumin during the growth period, crop density was adjusted at 120 plants m^{-2}

Cumin has relatively high seed growth rate (Fig. 3-3). However, seed filling period is too low and is highly influenced by day length and temperature (16, 21).

3-9 HARVEST INDEX

Cumin has a high harvest index (ratio of grain yield to total aboveground biomass at harvest) compared with many other crops. Small size of plant, narrow leaves, rapid allocation of dry matter during grain filling period, location of seeds at the uppermost layer of canopy and photosynthetic ability of seeds are the main causes of high harvest index of cumin. Reported values for harvest index of cumin (Table 3-5) are in the range of 0.43-0.61 and environmental conditions are the main source of this variation. Improvement of harvest index could be considered as an approach for increasing grain yield in many crops, but considering that cumin allocates more than half of its total aboveground biomass to grains, there is limited room for further increase in harvest index of this species.

Table 3-5 *Reported harvest index of cumin under different experimental conditions*

Biological yield (Kg ha^{-1})	Grain yield (Kg ha^{-1})	Harvest index	Experiment location	Reference
1,400	854	0.61	Mashhad (greenhouse)	17
1,600	945	0.59	Torbat-e-Jam	7
1,000	430	0.43	Feizabad-Mahvelat	14
1,228	517	0.44	Shiraz	1
1,300	666	0.51	Mashhad	13
2,317	1,357	0.59	Isfahan	2

3-10 GROWTH ANALYSIS

Growth is defined as increase in plant dry matter during time and, therefore, estimation of growth indices provides an analytical tool for description and interpretation of plant growth response to environmental conditions. Leaf area index (an index of plant photosynthetically active surface) and dry matter, when measured in given time intervals during growth season, are two basic variables from which several growth indices could be calculated (5).

3-10-1 Leaf Area Index

Leaf area index is a dimensionless index showing photosynthetic surface of plants, and its maximum value in many crops such as corn and wheat is around 5 (9). In cumin, stems and grains can also be considered as photosynthetic organs. Kafi (15), in a study on grain filling period of cumin, separated different photosynthetic organs and showed that contribution of leaves, stems and reproductive organs in plant photosynthetic surface were 35.6, 26.1 and 38.3%, respectively. Therefore, green area index (GAI) should be used instead of LAI in cumin growth analysis. Low potential for dry matter production in cumin seems to be partly related to its leaf area because GAI of cumin is low, rarely exceeding 1.5 and vegetative growth of plant at early stages is too slow, and the maximum GAI of cumin is obtained by the end of vegetative growth (Fig. 3-3), hence during the main part of growing season there is a limited photosynthetic surface for radiation absorption (Fig. 3-4). In an experiment at Torbat-e-Jam (23) during 75% of growth period LAI of cumin was lower than 0.2 and maximum LAI was reached 150 days after planting (10 days before harvest). In the study of Kafi and Rashed (16) maximum LAI of cumin was reported two weeks before harvest (89 days after planting) and LAI was below 0.2 in 60% of growth period (Fig. 3-4).

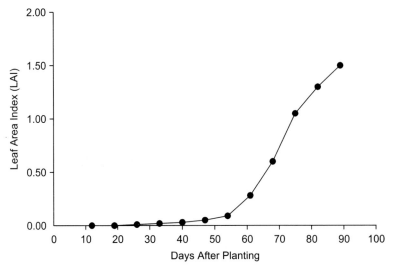

Fig. 3-4 Changes in leaf area index of cumin at crop density of 120 plants m^{-2}. Cumin was planted on March 4 and harvested 103 days after planting, last sample was taken 89 days after planting.

Low LAI during the season leads to low leaf area duration (LAD) and leaf area index duration (LAID). For example, LAID of cumin during the whole growth period is reported to be 29 m^2 leaf m^{-2} ground (Fig. 3-5) whereas LAID of corn hybrids is more than 330 m^2 leaf m-2 ground (1), which is about 15 times higher than that of cumin.

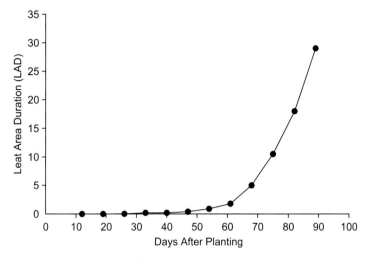

Fig. 3-5 Leaf area duration (m^2 leaf m^{-2} ground) of cumin at crop density of 120 plants m^{-2}. Cumin was planted on March 4 and harvested 103 days after planting, last sample was taken 89 days after planting.

3-10-2 Crop Growth Rate

While maximum crop growth rate (CGR) of species such as corn and wheat is higher than 30 g m^{-2} day^{-1} (16), CGR of cumin during the main part of growing season is lower than 1 g m-^{2} day^{-1} and reaches to its highest value of 6 g m^{-2} day^{-1} only in a short span of growth period (16, 23). In the early stages of growth, dry matter accumulation is too slow, leading to low competition ability against weeds (20). Experimental results have shown that more than 60% of total aboveground dry matter was produced during the last month of growth where CGR was at its maximum (Fig. 3-3).

3-10-3 Net Assimilation Rate

To the best of our knowledge, there is no published data on assimilation rate of cumin and its photosynthetic capacity is only estimated using net assimilation rate (NAR). Reported value for NAR of cumin in the early growth stages (28 days after planting) was about

10 g m^{-2} leaf day^{-1} (16, 23) and decreased later in the season (Fig. 3-6), but its reduction was low compared to many crop species due to low LAI, competition between leaves could be ignored.

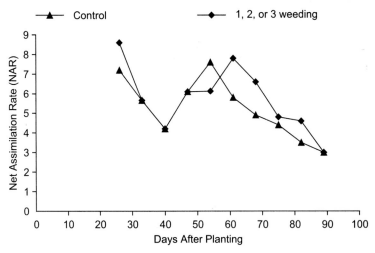

Fig. 3-6 Change in net assimilation rate (g m^{-2} leaf day^{-1}) under different weed control treatments. Since there was no significant difference between LAI in 1, 2, and 3 times weeding, these treatments were pooled.

3-10-4 Relative Growth Rate

Relative growth rate (RGR) indicates increase in dry matter in relation to initial biomass during a given time. In the early growth stages, RGR is high because most of the plant tissues are photosynthetic, but at later stages higher contribution of structural tissues in total dry matter and decreased rate of net assimilation as well as leaf senescence will lead to lowering the RGR (9, 23). Experimental results showed that four weeks after planting, RGR of cumin was 0.3 g g^{-1} day^{-1}, which was decreased to 0.04 g g^{-1} day^{-1} 13 weeks after planting (16). The main reason for this reduction in RGR is due to decrease in its components, i.e. net assimilation rate (NAR) and leaf area ratio (LAR) (Fig. 3-7).

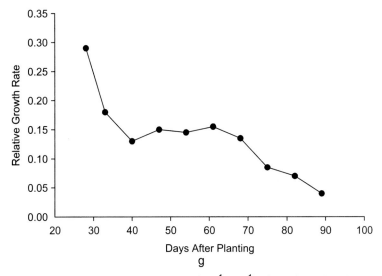

Fig 3-7 Changes in relative growth rate (g g^{-1} day^{-1}) of cumin during the growing season. Cumin was planted on March 4 and harvested 103 days after planting, last sample was taken 89 days after planting.

3-10-5   Specific Leaf Area and Specific Leaf Weight

Specific leaf area (SLA) is leaf area per unit dry matter and is an index for leaf thickness, e.g. the higher the SLA the thinner are the plant leaves. Salehi et al. (23) showed that SLA for cumin reached a maximum (0.03 m^2 g^{-1}) at flowering stage. In cumin, like many other crops, specific leaf weight (SLW), inverse of SLA, increases with plant age. Kafi and Rashed (16) showed that four weeks after planting, SLW of cumin was at its lowest (64.3 g m^{-2} leaf) and increased to a maximum value of 94.5 g m^{-2} leaf nine weeks after planting (Fig 3-8).

3-11   DEVELOPMENT STAGES

Crop development comprises two stages – (1) morphological, and (2) physiological. These events, during plant growth period, lead to transition from one functional stage to another. In grain plants, developmental processes usually start with seed germination and end

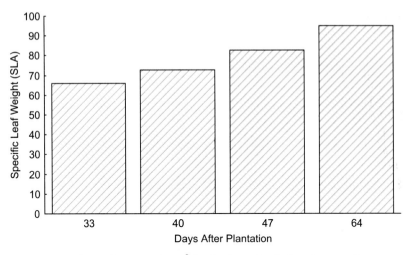

Fig. 3-8 Specific leaf weight (g m^{-2} leaf) of cumin during the growing season. Cumin was planted on March 4 and harvested 103 days after planting, last sample was taken 89 days after planting.

with seed maturity. While the development process in many crop species is fully distinguished and quantified (9), development stages of cumin is neither reasonably defined nor quantified.

Distinguished stages during cumin development include: seed germination, seedling emergence, appearance of first true leaf, appearance of leaf branching, stem formation (stems are usually double-branched but triple-branched stems may also be observed), umbrella formation, flowering, grain filling and maturity.

Heat units required for fulfilling developmental stage of cumin and duration of each stage is not fully understood. Aminpour and Mosavi (3) reported length of development stages under different irrigation treatments (Table 3-6).

Kafi and Rashed (16) calculated heat unit requirements of cumin assuming 6°C as the base temperature. Temperature data was collected from the weather station of Faculty of Agriculture, Ferdowsi University of Mashhad located 500 m from the experimental field. On this basis, heat requirement of cumin from planting to maturity was calculated as 1,007 degree-days. Generally, experimental results show that cumin development at early stages is very slow. However,

Table 3-6 *Duration of development stages of cumin (days after planting) under 1 (I1); 2 (I2); 3 (I3); and 4 time (I4) irrigation, seed was planted on December 17.*

Development stage	I1	I2	I3	I4
Emergence	43	43	43	43
3-leaf stage	63	63	63	63
5-leaf stage	78	78	78	78
Umbrella formation	117	117	117	117
Anthesis	133	133	133	133
Seed filling	140	140	143	143
50% maturity	151	151	158	163
80% maturity	155	155	163	168
Harvest	155	155	163	168

in the reproductive stage, development rate is too high. For example, results of Aminpour and Mosavi (3) showed that while duration of planting of cumin to five-leaf stage was 78 days, development from anthesis to maturity took only 30 days. It should be noticed that when seed is sown in late autumn or early winter where temperature is below germination base temperature, emergence might be delayed until March. Therefore, length of growth period of cumin is reported from 100 to 170 days (3, 15, 18, 19, 23).

In conclusion, it seems that the following development stages could be used for proper qualitative description of cumin phenology. Further research is needed to develop a quantitative model for cumin development based on temperature.

1. Emergence
2. Appearance of branched leaves
3. Stem formation
4. Flowering and seed formation (flowering process is not homogenous and flowers appear from the outermost part of umbrella inward)
5. Grain filling
6. Grain maturity

These development stages are shown in Appendix photographs 3-3 to 3-8.

3-12 RESPONSE OF CUMIN TO ENVIRONMENTAL STRESSES

In Iran and other cumin growing countries such as Egypt, Morocco and India, production areas are located in arid and semi-arid regions (19, 24) where drought, salinity, heat, cold, and wind prevail as single or multiple stresses (4). For centuries cumin has been able to grow successfully under these conditions and overcome abiotic environmental stresses. Short growth period of cumin is a response to environmental conditions in order to avoid heat and drought stresses of late spring. However, it does not mean that stress is not experienced by cumin during growing season.

3-12-1 Drought Stress

Cumin is adapted to drought and is grown as a dry land crop. In fact, in many parts of Iran cumin grows under rainfed conditions or by using partial irrigation practice. Experimental results have confirmed that cumin shows positive response to limited irrigation (3, 14) and there is no significant difference between yield of irrigated and rainfed crops, particularly in wet years (24). Furthermore, under rainfed conditions high spring rainfall may not always ensure a higher yield. In a two-year experiment with 151.6 and 317.1 mm rainfall during growing season, yield was higher in the first (drier) year (24), which was mainly due to prevalence of damping-off disease in the wet year. Aminpour and Mosavi (3) showed that while total rainfall during growth period was 78.8 mm, grain yield did not significantly differ between two and three times irrigation treatments. They recommended 1,637 m^3 ha^{-1} as water requirement of cumin for obtaining maximum grain yield. It seems that low transpiration surface and efficient water absorption capacity of cumin roots were the main reasons for this low water requirement. Further investigations are required to test this hypothesis.

3-12-2 Salt Stress

Salinity of soil and water could be regarded as important limiting factors in cumin growing regions. However, local farmers have different views about the response of cumin to salinity. While some producers insist that cumin could be irrigated with saline water, others believe that it would be possible only in wet years using moderately saline water (EC<5 dS m^{-1}). Generally, in light-textured soils, water with electrical conductivity of 6 dS m^{-1} could be used for irrigation.

Published information on salt tolerance of cumin is lacking. In controlled environment, germination of cumin seeds was studied under different levels of water potential using NaCl and Polyethyleneglycol (PEG) as osmoticum (Kafi, unpublished results). Both NaCl and PEG had negative effects on seed germination (Table 3-7). However, negative effects of PEG were more pronounced compared to NaCl indicating precautions needed in the use of PEG in germination tests on cumin.

Table 3-7 *Germination of cumin seeds in different water potentials using NaCl and PEG-6000.*

Type of osmoticum	Water potential (Mpa)	Germination (%)
Distilled water (control)	0.0	87.5
PEG	0.3	12.5
PEG	0.6	6.16
PEG	0.9	7.8
NaCl	0.3	63.3
NaCl	0.6	18.5
NaCl	0.9	13.3

Tawfik and Noga (25) used distilled water and PEG in different molecular weights as priming agent for cumin germination. Cumin seeds were exposed to different treatments for 1, 2, 3, and 4 days. Results showed that 100% germination of seeds and high vigor seedlings were obtained after 2-3 days exposure to distilled water or PEG-4000. They concluded that cumin seeds that are soaked for 3 days could be germinated satisfactorily until 30 days after priming.

In a sand culture experiment conducted in greenhouse, effects of NaCl and CaCl$_2$ at 0, 50, 100, 150, and 200 mol m^{-3} concentrations were studied on growth indices and yield of cumin (18). Results showed that yield and yield components were significantly reduced with increasing salinity levels but essential oil was not affected by salt stress (Table 3-8). An increase in Na$^+$ and a decrease in K$^+$ content of leaves and root of cumin was also reported in this experiment (Figs. 3-9 and 3-10).

Table 3-8 *Effects of different salt concentration in growing media on yield and yield components of cumin. Plants were grown in sand culture and irrigated with nutrient solution containing different salt levels.*

Salt level (mol m^{-3})	Grain yield (g m^{-2})	Biological yield (g m^{-2})	Number of umbrellas per plant	Number of seeds per plant	1,000-seed weight (g)	Essential weight
Control	113.42 a	183.64 a	15.0 a	17.0 a	2.80 a	2.95 a
50	95.97 b	165.12 a	13.0 a	13.0 b	2.50 b	2.72 a
100	46.47 c	103.12 b	7.0 b	10.0 c	1.40 b	2.51 a
150	32.40 c	76.47 cd	4.0 c	8.0 cd	1.10 c	2.27 a
200	19.98 d	51.55 d	1.0 d	6.0 d	0.80 d	2.14 a

a, b, c, d : In each column means following the same letter is not significantly different at 5% probability level.

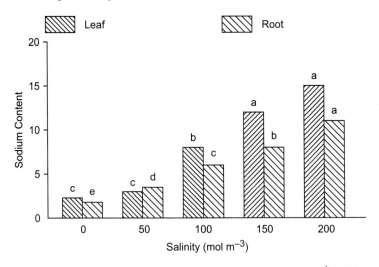

Fig. 3-9 Effect of different levels of salinity on Na$^+$ content (mg mg^{-1} DM) in leaves and root of cumin. a, b, c, d : Means following the same letter is not significantly different at 5% probability level.

In a pot experiment, cumin was irrigated using saline water with EC of 0, 4, 12, and 16 dS m^{-1} (26). In salinity levels higher than 8 dS m^{-1}, a significant reduction in yield, nutrient absorption and metabolite contents such as total chlorophyll, starch and soluble proteins was observed.

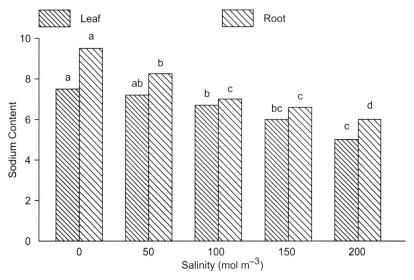

Fig. 3-10 Effect of different levels of salinity on K$^+$ content (mg mg^{-1} DM) in leaves and root of cumin. a, b, c, d : Means following the same letter is not significantly different at 5% probability level.

Zidan and Elewa (27) studied the effects of sodium chloride salinity at 0, 40, 80, 120, 160, 200, 240 and 280 mmols on germination, growth characteristics and carbohydrates, proteins and proline accumulation in cumin. Germination was severely reduced at higher NaCl levels. While the reasons were not explained, at low NaCl concentrations seedling growth was improved (Table 3-9). Plant water content was not affected by salinity and significant growth reduction was only observed in 200 mmols of NaCl. It seems in this experiment cumin exposure time to salinity was too short, therefore, salinity effects at lower NaCl concentrations were not visible.

Table 3-9 *Effect of different NaCl concentrations on stem and root length (cm), water content (100g⁻¹ fresh weight), and seedlings dry weight (mg) (27)*

NaCl (mmol)	0	40	80	120	160	200	LSD 1%
Stem length	8.40	9.40	10.80	9.50	7.20	6.50*	1.48
Root length	10.20	11.30	12.50	10.00	9.50	8.10	1.85
Water content	92.75	92.59	92.38	92.85	93.50	94.00	2.72
Dry weight	7.25	7.41	7.62	7.15	6.50	6.00*	1.20

* Significant difference with control

Soluble carbohydrates were increased with salinity level and this was more pronounced in intermediate NaCl concentrations. While plant protein remained unchanged at low salt levels, a moderate reduction in proline content was observed at higher NaCl concentrations, in addition total amino acids content of seedlings was reduced with salinity level (Table 3-10). Based on these results, they concluded that cumin could tolerate low to intermediate levels of salinity without considerable growth reduction.

Table 3-10 *Effect of different levels of sodium chloride on accumulation of carbohydrates (mg g⁻¹ DM), proteins (mg g⁻¹ DM), proline (mmol g⁻¹ DM) and other amino acids (mmol g⁻¹ DM) in cumin (27)*

NaCl (mmol)	0	40	80	120	160	200	LSD 1%
Carbohydrates	100.2	121.5 *	135.3	130.5 *	115.7 *	109.8 *	8.51
Proteins	130.4	133.2	131.2	130.8	120.1 *	110.7 *	6.31
Amino acids	20.9	18.8	16.4 *	16.2 *	15.8 *	15.1 *	2.72
Proline	7.6	8.2 *	8.9 *	9.1 *	9.5 *	10.3 *	1.51

* Significant difference with control

3-13 SUMMARY

Cumin is a temperate species; its growth period ends before hot summer temperatures and high relative humidity is the main limiting factor in cumin growing regions. Major yield components are number of plants per unit area, number of umbrellas per plant, number of seeds per umbrella and seed weight. While cumin can compensate for its yield components, for proper yield a good crop stand is necessary. Cumin has a high harvest index and seeds account for more than 50% of total above ground biomass. All aboveground organs besides petals, have photosynthetic activity, but green area index (GAI) of cumin is low and never exceeds 1.5. Low GAI in turn leads to low crop growth rate (CGR) and net assimilation rate (NAR). Cumin has a good resistance to drought and salinity and because of prevalence of wilting and blight disease under humid conditions, yield is rather higher in dry years. In many circumstances reasonable yield may be obtained with three times irrigation. Further investigations are required for better understanding of ecophysiological aspects of growth and development of this valuable species.

REFERENCES

1. Agraval, P. K. 1993. Handbook of Seed Testing. National Seeds Corporation Limited, New Delhi, India.
2. Alavi, A. 1970. Damping off disease of cumin. Journal of Iranian Society of Plant Disease, Agricultural Ministry. Pp. 92-98.
3. Aminpour, R. and S.F. Mosavi, 1995. Effects of irrigation intervals on development stages, yield and yield components of cumin. Journal of Agricultural Sciences and Natural Resources, 1: 1-7.
4. Basra, A.S. and R.K. Basra. 1997. Mechanisms of Environmental Stress Resistance in Plants. Taylor and Francis Publishers, UK, 432 p.
5. Bullock, D.G., R.L. Nielsen and W.E. Nyquest., 1998. A growth analysis comparison of corn grown in conventional and equidistance plant spacing. Crop Science, 28: 254-258.
6. Champawat, R.S. and V. N. Pathak. 1998. Role of nitrogen, phosphorous and potassium fertilizers and organic amendments in cumin (*Cuminum cyminum*) infected by *Fusarium oxysporum.* Indian Journal of Agricultural Sciences, 58(9): 728-730.
7. Chandula, R.P., S.C. Mathur and R.K. Strivastava. 1970. Cumin cultivation in Rajasthan. Indian Farming, July: 13-16.
8. Ehtramian, K. 2002. Effects of different nitrogen levels and planting dates on yield and yield components of cumin in Koshkak region, Fars province. M.Sc. Thesis, Shiraz University, Iran.

9. Evans, L.T. 1978. Crop Physiology. Cambridge University Press, UK.

10. Fotovat, A. 1993. Effects of macro-nutrients (N,P,K) on cumin yield. Scientific and Technological Research Organization of Iran, Khorasan Research Center, Iran.

11. Gary, B.K., B. Uday and S. Kathju. 2002. Responses of cumin to salt stress. Indian Journal of Plant Physiology, 7: 70-74.

12. Gora, D.R., N.L. Meena and P.L. Shiuran. 1996. Effect of weed control and time of nitrogen application in cumin (*Cuminum cyminum*). Indian Journal of Agronomy, 41: 500-501.

13. Hajian Shahri, M. 1997. Chemical control of cumin blight disease. Scientific and Industrial Research Organization of Iran, Khorasan Research Center, Iran.

14. Jangir, R.P. and R. Singii. 1996. Effect of irrigation and nitrogen on seed yield of cumin (*Cuminum cyminum*). Indian Journal of Agronomy, 41: 140-143.

15. Kafi, M. 1991. Studies on the effects of weed control intervals, row spacing, and plant density on growth and yield of cumin. MSc. thesis, Ferdowsi University of Mashhad, Iran.

16. Kafi, M. and M.H. Rashed Mohassel. 1993. Effects of weed control intervals, row spacing, and plant density on growth and yield of cumin. Journal of Agricultural Science and Technology, 6: 151-158.

17. Kafi, M., A., Zare Faizabadi and A.A. Mohammadabadi, 1994. Studies on the effects of seeding rate on irrigated and rainfed cumin. Scientific and Technological Research Organization of Iran, Khorasan Research Center, Iran.

18. Nabizadeh, M.R. 2003. Effects of different salinity levels on growth and yield of cumin. MSc. thesis, Ferdowsi University of Mashhad, Iran.

19. Patel, K.S., J.C. Patel, B.S. Patel and S.G. Sadaria. 1991. Water and nutrients management in cumin (*Cuminum cyminum*). Indian Journal of Agronomy, 36: 627-629.

20. Rahimi, M. 1994. Studies on chemical weed control in cumin fields. Scientific and Industrial Research Organization of Iran, Khorasan Research Center, Iran.

21. Rahimian Mashhadi, H. 1992. Effects of planting date and irrigation regimes on growth and yield of cumin. Scientific and Industrial Research Organization of Iran, Khorasan Research Center, Iran.

22. Raychaudhuri, S.P. 1992. Recent Advances in Medicinal, Aromatic and Spice Crops (Vol. 1). Today and Tomorrow's Printer Publication, New Delhi, India.

23. Salehi, M.R., M. Alikhordi, H. Ahmadi and Soleimairodi, H. 1997. Effects of nitrogen fertilizer levels on growth indices, yield and yield components of cumin in Torbat-Jam climatic conditions. Project Report, Agricultural Research Center of Khorasan, Iran.

24. Sadeghi, B. 1992. Effects of nitrogen levels and irrigation on cumin productivity. Scientific and Industrial Research Organization of Iran, Khorasan Research Center, Iran.

25. Tawfik, A. and A. Noga. 2001. Priming of cumin (*Cuminum cyminum*) seeds and its effects on germination, emergence and storability. Journal of Applied Botany, 75: 216-220.

26. Zargai, A. 1989. Medicinal Plants. Tehran University Press, Tehran, Iran.

27. Zidan, M.A. and M.A. Elewa. 1994. Effect of NaCl salinity on the rate of germination, seedling growth, and some metabolic changes in four species (umbeliferae). Der Tropen landwirt, Zeitsch-rift fur die land wirtschaft in den tropen und Subtropen, 95: 87-97.

Photo 3-1 Cumin field planted in 40 cm raw space

Photo 3-2 Cumin field planted in 20 cm raw space

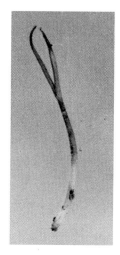

Photo 3-3 Cumin seedling emergence

Photo 3-4 Emergence of branch leaves in cumin

Photo 3-5 Cumin branching stage

Photo 3-6 Cumin flowering and seed setting stage

Photo 3-7 Cumin seed filling stage

Photo 3-8 Cumin seed ripening stage

CHAPTER 4

Production Technology for Cumin

A. Mollafilabi
filabi@kstp.ir
Iranian Research Organization for Science and Technology, Khorasan Center, Mashhad Iran

4-1   INTRODUCTION

Cumin is one of the important medicinal plants of India, Iran and many other countries and due to its specific ecological requirements it is grown only in a limited area of these countries (19, 24). About 90% of Iranian cumin designated for export is grown in Khorasan and the rest is produced in Kerman, Semnan, East Azerbaijan, Fars and Yazd provinces (14, 22, 24) Nearly the same ratio of Indian cumin is produced in Rajasthan. Cumin production is economically justified in water-limited areas and since this crop is mainly grown for export, farmers are assured of a stable and reasonable price (12, 14, 22, 24, 26).

4-2   CUMIN IN ROTATION

Cumin has been recommended in rotation with summer crops such as sorghum, corn, soybean, millet and sesame (9, 18, 24).

4-3 FERTILIZER APPLICATION

Nutrients are important components of crop growth and development. Since dry matter production in cumin, compared with other crops such as wheat, is much lower (two tons compared with 10 tons ha^{-1}, nutrient requirement for this crop is proportionally lower. However, nutrient requirement should be met by application of proper amount of fertilizers.

Application of 30 kg N, 60 kg P and 30 kg K per hectare has been demonstrated to produce satisfactory yield under environmental conditions of Mashhad (19). In Table 4-1 results of a fertilizer trial are shown.

Table 4-1   *Effects of different dose of nitrogen (N), phosphorus (P_2O_5) and potassium (K_2O) fertilizer application on seed yield of cumin (kg/ha). (19)*

K_2O (kg/ha)	30	30	30	0	0	0
P_2O_5 (kg/ha)	60	30	0	60	30	0
N (kg/ha) 90	478	498	434	328	534	349
60	523	483	499	466	435	442
30	597	420	418	317	339	370
5	440	161	417	309	454	455

In another experiment (14) with application of 30 kg N and 30 kg K per hectare, maximum yield was obtained. However, Sadeghi (24) found no difference in yield by application of 25, 50 or 100 Kg N per hectare. He stated that rainfed production of cumin is possible if adequate moisture is available at sowing (by end of January) with nearly 150 mm rainfall from February to end of April.

In such conditions yield up to one ton ha^{-1} could be produced (24). Javaheri (11) in a trial on the effects of plant density and nitrogen level in Jiroft area of Kerman province found that seeding rate of 14 kg/ha together with application of 50 kg N ha^{-1} was the best combination of these two factors. In this experiment seeding rates were 8, 14, 20 and 24 kg ha^{-1} combined with application of 0, 50, 100 and 150 kg N ha^{-1} (Fig. 4-1)(11).

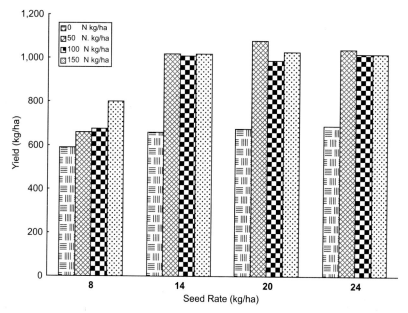

Fig 4-1 Interaction effects of different amount of nitrogen fertilizer (kg/ha) and seeding rate (kg/ha) on yield of cumin in Jiroft area, Iran (11)

Application of 150 kg ha^{-1} urea has been reported to produce satisfactory yield in Torbat-e-Jam (18). In an experiment in Koshkak (Fars province), with application of 0, 30 and 60 kg N/ha, it was found that with 30 kg N/ha maximum yield was obtained (7).

Niazi and Raja (20) found that application of 22.4 kg ha^{-1} N and 9.6 kg ha^{-1} P showed the highest yield. Fageria (8) reported differences in yield between 0-50 kg N and P ha^{-1} and 0 to 80 kg K ha^{-1}. Maximum yield was obtained with 50-50-80 NPK application. Sefa (25) found that application of 70 kg ha^{-1} N and P$_2$O$_5$ under a proper irrigation regime was the most economic fertilizer treatment. Champwat and Pathak (4) in India, evaluated effects of nitrogen, phosphorus, potassium and manure on damping-off disease and yield of cumin in a low fertility soil and noted that infestation was at the lowest, hence the maximum yield was obtained when 30-20-30 kg/ha NPK was applied. Chaudhary (6) found differences in response to application of a range of nil to 30 kg N ha^{-1} in combination with weed control and planting methods. Hornok (10) stated that a single application of top dressed nitrogen 30 days after sowing compared with split application (50% at sowing and 50% at 30 days after sowing) caused yield increase of 19 and 12% respectively. It was shown that application of 9 kg ha^{-1} P increased yield by 9%.

In general, for cumin 15–20 tons ha^{-1} manure together with phosphorus and application of nitrogen at a rate of 25–30 kg ha^{-1} , 40–50 days after sowing is recommended.

4-4 PLANTING

4-4-1 Seedbed Preparation

Seedbed preparation is crucial for proper plant establishment (15). For seedbed preparation, after application of a suitable amount of manure, the soil is plowed in early autumn. Since seeds of cumin are small, a firm and smooth bed is required in order to ensure adequate contact of seeds with soil and, therefore, better seed germination and plant establishment. Seeds are normally sown by traditional methods but in recent years by modification of cereal sowing equipment, row seeding for cumin is practiced.

Cumin seed is sown in a dry or moist bed in the form of row planting or flat plots. This crop is normally produced as rainfed but in recent years irrigation is also practiced. When sowing is practiced in day beds, after plowing and seedbed preparation, seeds are sown. In moist bed sowing system, after irrigation of soil and when moisture content reaches field capacity level, seeds are sown and covered by soil through a light disc or cultivator (1, 26) . Broadcasting of seeds is a normal practice but row seeding is also possible when modified cereal drill is used. Seeds are broadcasted by hand or seeding equipments and furrows 40-50 cm apart are made by a furrow opener. Sowing sequences may differ in different parts of the world based on availability of equipments and local experiences. The normal practice is based on plowing–disking–sowing–light disking–wooden leveler.

Row planting with furrow irrigation system has been found to perform satisfactorily in Torbat-e-Jam (17). However, in India (5, 6) methods of planting have not shown any difference in yield performance.

4-4-2 Date of Sowing

Date of sowing is an important factor in plant growth and development (26). In an experiment under climatic conditions of Mashhad (23) with four planting dates (10 December, 1 January, 5 March, and 25 March), early planting (10 December and 1 January) showed better results (Fig. 4-6).

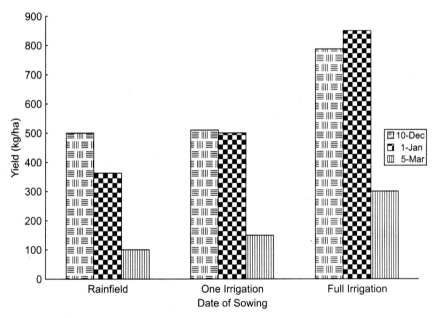

Fig. 4-2 Effect of four date of sowing and three irrigation regime on seed yield of cumin in Mashhad, Iran (23)

In this experiment, reduction in number of umbels per plant has been found to be associated with late planting due to high sensitivity of plants to day length (23). Mollafilabi (16) found that late December to early January sowing was superior to late February/early March for two areas of Ghaen and Torbat-e-Jam.

However, in Koshkak area of Fars province, Iran, it was found that planting in early February and early March is preferred (7) (Fig. 4-3). This was related to risk of freezing which may occur in the region.

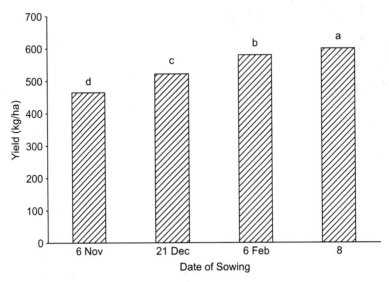

Fig. 4-3   Effects of four dates of sowing on seed yield of cumin, Fars province of Iran (7)

In India, date of planting of cumin varies from 15 November to late December (10). Early planting provides enough time for the plant to develop before entering long day periods.

It is, therefore, concluded that in warmer areas where winter freezing is not a problem, early planting in December and in cold areas late planting (February) is recommended.

Early planting has resulted in better yields (16, 22, 23).

4-4-3   Sowing Density

The amount of seeds required for sowing depends on soil type, moisture and fertility, methods of planting, sowing date, seed viability, and also farming practices. Kafi (12) found that 1,200,000 plants/ha with rows 40 cm apart gave the best result (Fig. 4-4).

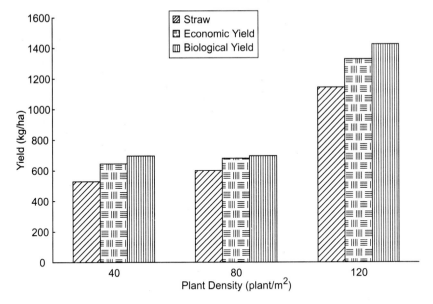

Fig. 4-4 Biological and seed and straw yield of cumin under three plant densities (12).

Mollafilabi (16) in an experiment on the effects of planting date and density under rainfed and irrigation systems, found that generally there was not much difference in row spacing of 50, 60 or 70 cm. However, in late planting, an interaction effect was found. In this case narrow rows gave better results only in late planting in February – early March

In another trial, Mollafilabi (17) found no significant difference in yield between different seeding rates (5, 10, 15, 20, 25 and 30 kg/ha) and two irrigation systems (Table 4-2).

Kafi (13) also found no differences in yield between a wide range of seeding rates of 4, 8, 12, 16 and 20 kg ha[-1] under both rainfed and irrigated systems (Figs. 4-6 and 4-7).

Seeding rate of 14 kg ha[-1] has been recommended for Jiroft area of Kerman province (11). Usually, in India 12 kg/ha is used, but in Rajasthan (India), density of 1,330,000 plants ha[-1] is practiced (12). Considering density of 1,200,000 plants ha[-1] recommended earlier (12) and taking into account the seed viability of 80% and 1,000-

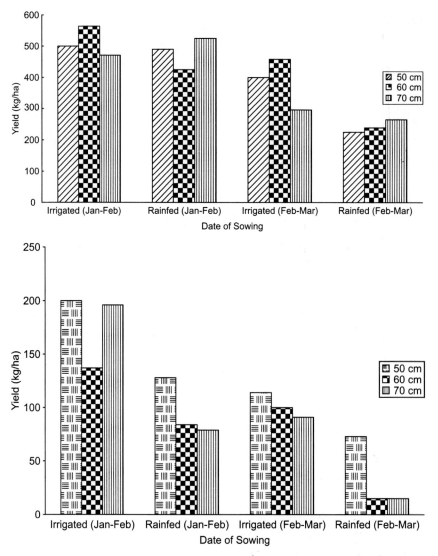

Fig. 4-5   Effects of row spacing, planting date and irrigation regime (Rainfed, irrigated) in two Khorasan regions (A. Torbat-e-Jam, B- Ghaen) (16)

seed weight of 3.2-3.8 g, it appears that 5 kg ha^{-1} seed rate is an appropriate amount. However, under traditional systems a much higher amount is used, e.g. in Torbat-e-Jam, Khaf, Torbat-e-Hydariah 15 kg, in Faizabad 25 kg ha^{-1}, and in Khomain 30 kg ha^{-1} is

Table 4-2 *Seed yield of cumin with different seeding rates and irrigation systems. Similar letter beside each data in third column of different cells shows no significant differences between treatments (17)*

Seeding rates	Irrigation systems	Yield
5	Furrow irrigation	768 abc
10		735 bc
15		774 bc
20		732 bc
25		707 bc
30		789 abc
5	Flood irrigation	663 c
10		860 abc
15		780 abc
20		820 abc
25		1,022 abc
30		964 ab

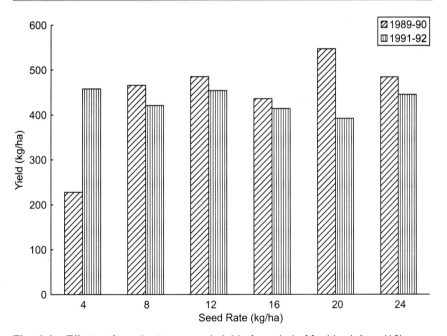

Fig. 4-6 Effects of seed rate on seed yield of cumin in Mashhad, Iran (13).

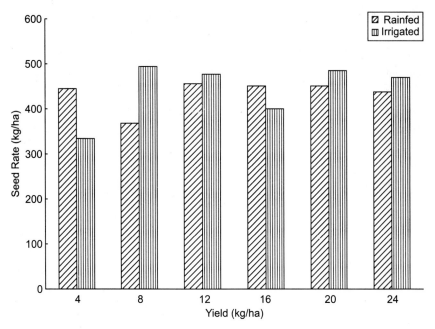

Fig. 4-7 Effects of seed rate on seed yield of cumin under rainfed and irrigated regime in Faizabad, Khorasan, Iran (13).

used. Higher rates are normally used in poor soils (e.g. in Faizabad up to 60 kg ha^{-1} seed is sown in sandy soils in the fringe area of deserts).

However, since cumin is very susceptible to fungal diseases, the sowing depth is 1-2 cm and deep sowing causes low seeding emergence and poor stand (21).

4-4-4 Seed Conditioning Prior to Sowing

It is recommended that seeds be soaked in slow running water for 24-36 hours prior to sowing for probable removal of any inhibitory compound (10). Where there is a risk of contamination with fungal diseases such as *Alternaria* or *Fusarium*, it may be necessary to disinfect the seeds with appropriate fungicides (22). Dense planting may cause further infection of theses diseases due to enhanced humidity in the canopy.

4-5 IRRIGATION

Cumin crop may be irrigated several times (up to six times) during growth period, based on availability of moisture.

However, a single irrigation has been shown (24) to be effective only when the amount of rainfall in the growth period is only 150 mm. Sowing in late winter has been shown (16) to require two to three irrigations, whereas early sowing may not require extra irrigation. In an experiment in Isfahan (2), it was found that the effect of irrigation was more pronounced at the time of pollination to seed filling. It has been stated (24) that in cases when rainfall coincides with plant growth, one irrigation at the time of flowering is effective.

4-6 HARVESTING

4-6-1 Seed Yield

Many factors affect seed yield of cumin and, therefore, the range of yield varies from a very small amount up to 2,000 kg/ha depending on the environmental factors affecting yield (3). Average yield of cumin under rainfed and irrigation conditions in Iran has been reported to vary from 100 to 1,000 kg/ha (17).

In a survey conducted for cumin in Iran (3), the following conclusions were made

1. This crop is mainly grown in Khorasan province
2. Other provinces of the country with minor production are Gholestan, Semnan, Central Yazd and Azerbaijan
3. Seed rate varies from 5-35 kg/ha mainly in the range of 10-15 kg/ha
4. Time of sowing varies from mid November to late February-mid January
5. Fertilizers used are ammonium phosphate (in the range of 100 to 300 kg/ha) and urea (in the range of 50-150 kg ha^{-1})

6. Number of irrigation is normally two times, but in some places it may be up to six times
7. Weeding once or twice during the growth period
8. Diseases and pests, *Alternaria*, *Fusarium*, Trips and ants
9. Time of harvest, late May to late June or early July
10. Yield ranges from 200 to 2,000 kg ha^{-1}.

4-6-2 Time of Harvest

Cumin seed is normally harvested from late May to late June in cold to moderate areas (i. e. Iran) and in January to February in subtropics and tropic areas (i. e. India). In the traditional system of harvesting, plants are cut with hand tools before shedding starts and seeds are separated from the straw by various means. However, in recent years attempts have been made to modify seed harvesters for this purpose (22). After cleaning, cumin seeds are kept in bags for further use. High-grade seed lots for export are not permitted to contain foreign materials.

4-7 Summary

Cumin is an important medicinal plant exported from Iran, and it is grown as a rainfed or irrigated crop mainly in Khorasan province. Seedbed preparation is important in cumin production due to its small seed size. The best time of sowing in Khorasan is December and the most common rate of sowing is 5 kg/ha with yield ranging from 200 to 2,000 kg/ha. Seeds may require disinfection against fungal diseases such as *Alternaria*.

REFERENCES

1. Aminotojari, A. 1993. Enhancing quality of cumin for export. Institute of Standard and Industrial Research, Khorasan, Iran.
2. Aminpour, R. and S. F. Moosavi. 1994. Effects of frequency of irrigation on growth development and yield components of cumin. Journal of

Science of Agricultural and Natural Resource, Isfahan University of Technology. 1: 1-7.

3. Balandari, A. 1994. Botanical characteristics of local population of cumin in Iran. Scientific report, Iranian Scientific and Industrial Research Organization, Khorasan Center.

4. Champwat, R. S. and V. N. Pathak. 1988. Role of nitrogen, phosphorus and potassium fertilizers and organic amendments in cumin (*Cuminum cyminum* L.). Indian. J. Agric. Sci. 58(9): 728-730.

5. Chowdhury, G. R. and O. P. Gupta. 1982. Effect of weed control, sowing method and nitrogen application on growth and quality of cumin (*Cuminum cyminum* L.) . Haryana. Agron. J. 5: 79-81.

6. Chowdhurry, G. R. 1989. Effect of nitrogen level and weed control on weed competition, nutrient uptake and quality of cumin (*Cuminum cyminum* L.). Indian J. of Agric. Sci. 59(6): 397-399.

7. Ehteramian, K. 2002. Effects of nitrogen level and date of sowing on yield and yield components of cumin in Koshkak area of Fars province. M. Sc thesis, Shiraz University.

8. Fageria, N. K. 1972. Effect of nitrogen, phosphorus and potassium fertilization on yield and yield attributing of cumin crop (*Cuminum cyminum* L.) Kreuz-Kummel, Zeitschrift Furpflanz Enernahruag and Bodenkund, 132(1): 30-34.

9. Fotovat, A. 1992. Effect of N, P, K nutrient on yield of cumin. Scientific report, Iranian Scientific and Industrial Research Organization, Khorasan Center.

10. Hornok, L. 1992. Cultivation and processing of medicinal plants. Academic Kiado, Budapest. 45: 26-77.

11. Javaheri, A. 1999. Effects of plant density and nitrogen rate on growth and yield of cumin in Jiroft area of Kerman province. M. Sc thesis, Islamic Azad University of Jiroft.

12. Kafi, M. 1990. Effects of frequency of weeding, row spacing and plant density on growth and yield of cumin. M.Sc. thesis, Faculty of Agriculture, Ferdowsi University of Mashhad.

13. Kafi, M. 1993. Effects of seeding rate on yield of cumin under rainfed and irrigated conditions. Iranian Scientific and Industrial Research Organization , Khorasan Center.

14. Karimi, P. 1989. Chemical analysis of essential oil of umbelifereae family, Tabriz University, Thesis of Doctor of Pharmacy, Iran.

15. Khajehpour, M. R. 1986. Principle of Crop Production. Jehad-e-Daneshgahi of Isfahan University of Technology.

16. Mollafilabi, A. 1992. Effects of date of sowing and row spacing on yield of cumin. Scientific report, Iranian Scientific and Industrial Research Organization , Khorasan Center.

17. Mollafilabi, A. 1993. Effects of rate of seeding and planting methods on yield of cumin. Iranian Scientific and Industrial Research Organization, Khorasan Center.

18. Mollafilabi, A. 1998. Effects of rate of nitrogen on physiological indices of growth and yield components of cumin. Proceedings of 5th Iranian Crop Science Congress, Karaj, Iran.

19. Naseri Pouryazdi, M. T. 1991. Effects of NPK on growth and yield of cumin. M.Sc. thesis, Ferdowsi University of Tarbiat Modarres.

20. Niazi, M. H. K. and M. R. Raja. 1971. Effect of NPK on yield of white Zeera (*Cuminum cyminum* L.) J. Agric. Research 9(2): 124-127.

21. Omidbigi, R. 1999. Approaches for production and processing of medicinal plants.Vol. 3. Behnashr Publishing Company, Astan Ghods Razavi, Iran.

22. Rahimi, M. 1993. Chemical control of weeds in cumin. Scientific report, Iranian Scientific and Industrial Research Organization, Khorasan Center.

23. Rahimian Mashhadi, H. 1991. Effects of date of sowing and irrigation on growth and yield of cumin. Scientific report, Iranian Scientific and Industrial Research Organization, Khorasan Center.

24. Sadeghi, B. and M. H. Rashed Mohasel. 1993. Effects of rate of nitrogen and irrigation on yield of cumin. Scientific report, Iranian Scientific and Industrial Research Organization, Khorasan Center.

25. Sefa, S. 1986. Nitrogen and phosphorus requirement of cumin growth under dry and irrigated conditions in Eskishir province. Hort. Abs. 58:331.

26. Zare Faizabadi, A. 1994. Cumin in Khorasan. Annual seminar, Faculty of Agriculture, Ferdowsi University of Mashhad.

CHAPTER 5

Irrigation of Cumin

A. Alizadeh
alizadeh@ferdowsi.um.ac.ir
Faculty of Agriculture, Ferdowsi University of Mashhad, Iran

5-1 INTRODUCTION

Information on water requirement of cumin is scarce and not much data has been documented on this issue. This may be due to the fact that in Iran cumin is not regarded as a main crop in cropping systems and it is grown as a supplementary crop. Also, cumin is mainly grown as a rainfed crop and irrigation is only practiced when excess water is available as supplementary irrigation. Therefore, the amount of water and time of irrigation is based on the availability of water. However, during the last few years, irrigated cultivation of cumin is being practiced and the trend is gradually changing from a rainfed to an irrigated crop.

Indigenous knowledge of farmers indicates that in the main cumin growing areas of central and southern Khorasan (Iran), there is no need for irrigating cumin and excess water may cause yield reduction due to prevalence of diseases such as *Alternaria*, to which cumin is susceptible.

5-2   RESISTANCE OF CUMIN TO WATER DEFICIT

Rainfed production of cumin is a good indication of resistance of cumin to water shortage. Shape of leaves, low plant height and color of leaves are all indices of good adaptation of cumin to dry areas (6). However, to evaluate the extent to which this plant responds to water shortage, proper investigations are required (2). In an experiment (14) the trend of changes in water potential under irrigated and rainfed conditions was evaluated in Mashhad (Iran) (Fig. 5-1). Maximum water potential was recorded before sunrise and the values for 3rd May, which coincides with maximum growth of cumin, was -4 bar for full irrigation and -7 bar for rainfed crop during the day. With evaporation increasing steadily, water potential was reduced and reached its minimum level at 1400 hours. The slope of the curve for decreasing water potential with time was even and similar for both systems, which indicates that physiological response of cumin was not affected by irrigation, and for controlling transpiration plants tend to close their stomata. Closing of stomata in full irrigation occurred under higher water potential compared with rainfed treatment (-14 bar in irrigated and -21.5 bar in rainfed crop).

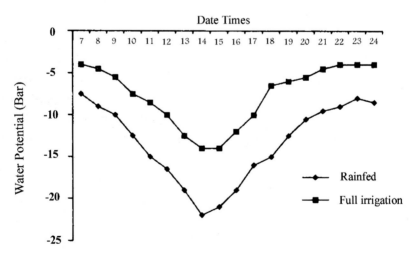

Fig. 5-1   Trend of changes in plant water potential during the day under rainfed and irrigation conditions (5).

It was shown (14) that water potential of cumin during growth period under full irrigation started from -4 bar and reached a maximum of -15 bar with a linear change, whereas under rainfed conditions these values ranged from -12 to -30 bar (Fig. 5-2).

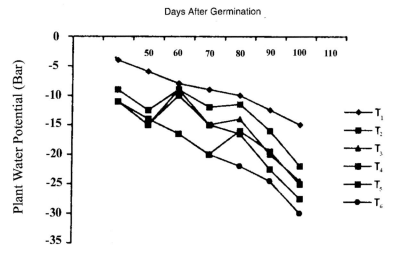

Fig. 5-2 Trend of plant water potential during the growth period (5)
T1 = rainfed, T2 = one irrigation at flowering, T3 = one irrigation at seed filling, T4 = T2+T3, T5 = one irrigation after seed emergence + T2+T3, T6 = full irrigation.

As shown in Fig. 5-2, changes in water potential with full irrigation and rainfed conditions are in a stable state with a linear form, whereas in other treatments with partial irrigation, water potential is fluctuating. In general, as the plant matures, water potential tends to decrease and the rate of recovery to the previous state due to further irrigation decreases. Drought tolerance of cumin has been associated with the ability of roots to absorb moisture from the soil, but other findings indicate that root extension of cumin is limited (1, 2).

In a study conducted in Isfahan (Iran) (1), it was reported that maximum length of the main root was 15 cm. Behboudi (4) also showed that the length of the main root increased linearly from the commencement of growth and reached a maximum of 12 cm, 70 days after seed germination, with not much changes towards the end of the growth period. He observed that the maximum length of the root coincided with 50 mm accumulative evapotranspiration. It is, therefore, concluded that this plant extracts water from the upper layers of soil in which moisture accumulates from frequent light showers. Therefore, well distributed light showers are more effective in cumin growth and development compared with heavy rains. Enough moisture during germination as well as in seed filling stage is crucial.

Rainfed farmers prefer light soils for cumin because moisture penetration in these types of soil during winter and spring is faster and, therefore, enough water is provided for the plant growth.

5-3   CROP COEFFICIENTS

Irrigation requirements of a plant are normally determined by crop coefficients during the growth period and if crop coefficient (Kc) and evapotranspiration for reference plant (ETo) are known, crop evapotranspiration (Etc) can be calculated as Etc=Kc. ETo (8).

Since Kc is not constant, in order to determine ETc in a specific period of time, it is necessary to draw the curve for fluctuation of this parameter during the growth period. FAO (1, 5) has proposed a method for evaluation of ETo and for this purpose crop growth period has been divided into four following stages.

5-3-1   Initial Stage

This stage commences from germination to a point where 10% of soil cover is achieved, which has been reported (4, 14) to be after 75 days. The average value of crop coefficient for this stage (Kc-ini) obtained from lysimeter and compared with evapotranspiration of a grass cover as a reference crop has been shown to be 0.34 (4).

5-3-2   Development Stage

This period starts from crop establishment up to a point where plant cover is 70-80%. For cumin, this period is 30 days and the mean crop coefficient is 0.52.

5-3-3 Mid-season Stage

In this stage plants are still physiologically active without any development and it lasts for 40 days with an average of 0.77 for Kc (Kc-mid).

5-3-4 End-season Stage

This is a stage in which the plant does not require water. Since cumin seed is sensitive to shattering, harvesting is conducted before this stage while plants are still green, otherwise most crops are harvested in this stage. This stage lasts 25 days for cumin with Kc (mid) of 0.43 (4). Table 5-1 shows growth periods and average Kc for cumin.

Table 5-1 *Growth period and crop coefficient for cumin (2).*

Growth stage	Length (days)	Kc
Initial stage	75	0.34
Development stage	30	0.52
Mid-season	40	0.77
End-season	25	0.43

In Fig. 5-3 crop coefficient curve for different growth periods of cumin is demonstrated.

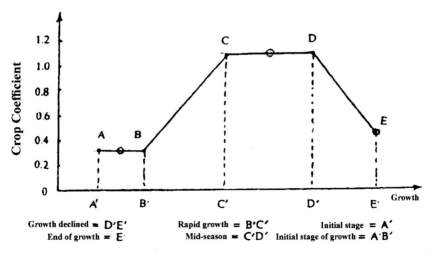

Fig. 5-3 Schematic curve for crop coefficient during growth period (6)

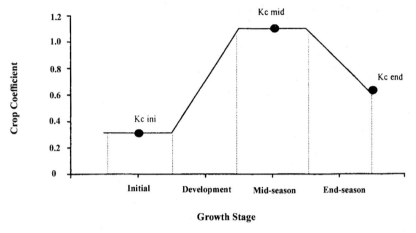

Fig. 5-4   Schematic curve for crop coefficient.

Based on an experiment conducted from 1998 to 2000 in Mashhad (Iran), the following Kc curve was obtained: Kc value for different crops in initial, mid and end season stages is presented in irrigation bulletin No. 56 of FAO. However, no value is shown for cumin. It has been recommended to make corrections in the Kc value based on moisture availability and wind speed in mid and end season. For this purpose the following formulae have been used:

(5-2) Kc -mid = Kc-mid (table) + [0.04 (U2 -2)-0.004 (RH_{min}– 45)] $[h/3]^{0.3}$

(5-3) Kc -end = Kc-end (table) + [0.04 (U2 -2)-0.004 (RH_{min}– 45)] $[h/3]^{0.3}$

Where:

Kc = mid is corrected crop coefficient in mid-season

Kc = mid-table is Kc in FAO table for which a value of 0.77 for cumin should be considered

Kc = end is corrected Kc at the end of season

Kc-end-table = Kc in FAO table for which a value of 0.43 should be considered

U2 = wind speed at a height of 2 cm above the ground level (cm/sec)

RH_{min} = minimum relative humidity (%)

h = plant height (m) which is 0.3 for cumin.

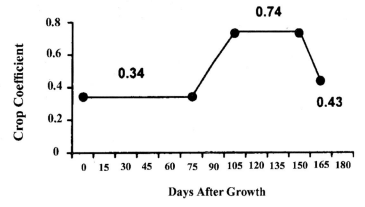

Fig. 5-5 Crop coefficient based on FAO (9)

5-4 WATER REQUIREMENT OF CUMIN

Water requirement (ETc) is the amount of potential water required for plant growth and evapotranspiration demand during the growth period. There is a lack of data for cumin in this respect. Behboudi (4) estimated potential ETo for cumin during the growth period equal to 560 mm and the water requirement equal to 335 mm or 3350 m³/ha. In other words, the average crop coefficient for the growth period is 0.59. However, these values have been estimated in lysimeter experiments in a full crop cover under climatic condition of Mashhad.

5-5 IRRIGATION REGIME FOR CUMIN

Irrigation regime is the management of the quantity and time of irrigation application. In experiments conducted in India in 1988 and 1989, quantity of 100, 150 and 200 mm was applied twice, three or four times during the growth period. It was shown that the highest yield was obtained with 200 mm water applied for times and this was attributed to higher number of umbels per plant and 1,000-seed weight (10, 11).

In similar experiments (7), application of different amounts of Nitrogen (0, 20, 40 and 60 kg/ha) together with four, five or six times irrigation was investigated. Results showed that by five or six times irrigation the highest yield was recorded and application of 20 kg/ha Nitrogen was sufficient for this crop. By increasing the number of irrigations, incidence of fungal diseases also increased. The highest amount of water applied in six time irrigation was 162 mm. However, in these experiments the amount of rainfall is not defined.

Rahimian (12) found that full irrigation showed the best result. Mollafilabi (9) found that twice irrigation was enough in Ghaen, and in Torbat-e-Jam a rainfed crop preformed satisfactorily. Sadeghi (13), in his experiments, found that application of excess water is not only beneficial, but may also cause harm due to incidence of diseases.

He concluded that irrigation is only beneficial when spring is dry and during the growth period total rainfall is less than 150 mm. In general, he recommended that twice irrigation at flowering and seed filling is enough for cumin. Aminpour and Moosavi (3), under climatic condition of Felavarjan in Isfahan, recorded yield of 923, 965, 1,742 and 1,800 kg/ha for one irrigation after sowing (I_1), one irrigation after sowing plus one at the time of crop establishment (I_2), T_1+T_2+one irrigation at onset of flowering (I_3) and $T_1+T_2+T_3+$one irrigation at seed filling (I_4) respectively. An increase in number of umbels per plant was the contributing factor in yield enhancement for I_3 and I_4 treatments. However, the number of seeds/umbel and 1,000-seed weight in these two treatments were lower than others.

Tavoosi (14), in a similar experiment, applied different amounts of water at different growth stages:

T_1 = rainfed

T_2 = one irrigation at flowering

T_3 = one irrigation at filling

$T_4 = T_2 + T_3$

T_5 = one irrigation after seed emergence + $T_2 + T_3$

T_6 = full irrigation

Under this condition the amount of rainfall was 190 mm and the amount of water received for the full irrigation was 350 mm. He noted that there was no difference between treatments in terms of yield, number of umbels and number of seeds/plant, but fully irrigated plots showed the lowest 1,000-seed weight and harvest index and also the highest biomass. It was concluded that in the years with a minimum of 160 mm rainfall there is no need for further irrigation. Behboudi (4) found that under lysimeter experiment, cumin utilized 240 mm water but yield was not necessarily at the maximum level.

5-6 EFFECTS OF IRRIGATION ON YIELD COMPONENTS

5-6-1 1,000 Seed Weight

Irrigation has been shown to decrease 1,000-seed weight (14). This has been associated with more vegetative growth and, hence, lowers allocation of nutrients to the seeds. The weight of 1,000 seeds was reported to be 2.17 g with full irrigation and 2.67 g under rainfed conditions (14). In other experiment (3), 1,000-seed weight was 3.03 for one irrigation and 3.27 g for twice irrigated plots. Three times or four times irrigation reduced 1,000-seed weight.

5-6-2 Number of Umbels Per Plant

Irrigation has shown to have negligible effects on number of umbels per plant, and under rainfed conditions this component was higher compared with irrigated plots (7). Application of five times irrigation increased the number of umbels per plant (7), but six times showed no effect. The amount of water applied with six times was 165 mm. Tavoosi (14) showed that the number of umbels per plant was 27.6 for full irrigation and 27.8 for rainfed crops. In another experiment (3), the number of umbels per plant was 19.9 for one irrigation and 19.2 for twice irrigation. In this experiment, application of higher amount of irrigation (three or four times) increased the number of umbels significantly (25.5 and 55.1 respectively).

5-6-3 Number of Seeds Per Umbel

Although there is evidence (3) which indicates that the number of irrigation reduces the number of seeds per umbel (30.3 in one irrigation to 22.6 in 4 irrigations), another reference (14) illustrates that there was difference in this respect between rainfed and full irrigation. An increase in number of seeds per umbel with increasing irrigation has also been found elsewhere (10).

5-7 BIOLOGICAL YIELD

Increasing the irrigation number has been found (14) to increase biological yield. Tavoosi (14) found that there was a significant difference between fully irrigated and rainfed, but no differences were observed between supplementary and full irrigation in terms of biological yield. Another evidence (3) also suggested no differences between once or twice irrigation, but there was a significant difference between three and four times irrigation .

5-8 ECONOMIC YIELD

It is generally believed that there is no need for irrigation of cumin in normal years and water may cause yield reduction. Aminpour and Moosavi (3) found that seed yield was 992.8 kg/ha with one irrigation and 1,800.3 kg/ha with four irrigations, whereas, Tavoosi (14) and Behboudi found no effect of irrigation on economic yield of cumin.

5-9 SUMMARY

In general, cumin is adapted to dry conditions and since its growth period coincides with winter and spring rainfalls, there is no need for extra irrigation. However, where there are prolonged dry periods, supplementary irrigation is effective. Crop coefficients for cumin in initial, mid-season and end-season are 0.34, 0.77 and 0.43 respectively with 3350 M^3/ha water requirement. Excess water may reduce yield due to spread of fungal diseases.

REFERENCES

1. Allen, R. G., L. S. Pereira, D. Raes and M. Smith. 1988. Crop evapotranspiration: Guidelines for computing crop water requirement. FAO Irrig. and Drain. Paper 56. Rome, Italy. 301 PP.

2. Alizadeh, A. 2001. Potential evapotranspiration and crop coefficient for cumin and saffron. Technical report, Iranian Scientific and Industrial Organization, Khorasan Center.

3. Aminpour, R. and S. F. Moosavi. 1997. Effects of number of irrigation on growth , yield and yield components of cumin. J. of Agriculture and Natural Resource Sciences , 1: 1-7.

4. Behboudi, S. 2001. Variability of crop coefficient for cumin under climatic conditions of Mashhad. M. Sc. Thesis, Ferdowsi University of Mashhad.

5. Doorenbos, J. and W. O. Pruitt. 1977. Crop water requirements. FAO Irrig. and Drain. Paper 24. Rome, Italy.

6. Duncan, W. G. 1980. Physiology of maize. In Evans (ed), Crop Physiology. Cambridge University Press, 374 pp, England.

7. Jangir, R. P. and Z. Singh. 1996. Effect of irrigation and nitrogen on seed yield of cumin. Indian J. Agron. 41: 140-143.

8. Keshavarz, A., and K. Sadeghzadeh. 2000. Optimal reduced irrigation rate: economic and mathematical analysis. J. of Technical and Engineering Research for Agriculture, 5: 1-26.

9. Mollafilabi, A. 1997. Effect of date of sowing and row width on yield of cumin under rainfed and irrigation. Technical report, Iranian Scientific and Industrial Organization, Khorasan Center.

10. Patel, K. S., J. C. Patel, B. S. Patel and S. G. Sadaria. 1991. Water and nutrient management in cumin. Indian J. Agron. 36: 627-629.

11. Patel, K. S., J. C. Patel, B. S. Patel and S. G. Sadaria. 1992. Influence of irrigation, nitrogen and phosphorus on consumptive use of water, water use and water expense efficiency of cumin. Indian J. Agron. 37: 209-211.

12. Rahimian, 1990. Effect of planting date and irrigation regime on growth and yield of cumin. Technical report, Iranian Scientific and Industrial organization, Khorasan Center.

13. Sadeghi, B. 1991. Effect of rate of N application and irrigation on cumin. Technical report. Iranian Scientific and Industrial Organization, Khorasan Center.

14. Tavoosi, N. 2000. Effect of irrigation regime on yield and yield components of cumin. M. Sc. Thesis, Ferdowsi University of Mashhad.

CHAPTER 6

Diseases, Pests and Weeds

M. Hagian Shahri[1] and M.H. Rashed Mohassel[2]
1 mhag52570@yahoo.com
Agriculture and Natural Research Center of Khorasan, Iran
2 mhrashed@ferdowsi.um.ac.ir
Faculty of Agriculture, Ferdowsi University of Mashhad, Iran

6-1   DISEASES OF CUMIN

One of the main drawbacks of cumin, specially during the rainy season, is the infection of cumin by fungal diseases. So far, three serious fungal diseases have been reported for cumin (33), *Fusarium* wilt of cumin *(Fusarium oxysporum)*, cumin leaf blight *(Alternaria burnsii)* and powdery mildew *(Erysipe polygoni)*, among which cumin wilt and leaf blight are very common in India and Iran (16, 17).

6-1-1   Fusarium Wilts *(Fusarium oxyosporum)*

The first investigation of cumin fungal disease was conducted by Gaur (1949) in India and he suggested that *Fusarium* spp. is the cause of infection (10). Further studies conducted by Mathur and Mathur (1965), identified *F. oxysporum* as the cause of wilting of cumin (27).

Patel and Prasad identify the cause of this disease as a physiological form of *F. oxysporum*. They reported the average intensity of this disease as 26% with high intensity up to 96% in Gujarat, India (39). They showed that when isolates of this disease agent are cultured in natural and artificial medium with different amounts of carbon and nitrogen sources, the morphological and physiological differences among isolates could be seen. The optimum growth for isolates also differed (32, 28).

Summer fallow reduces the infestation. On the other hand, planting cumin after sorghum increases cumin infestation to *Fusarium* wilt (34, 35). Intercropping cumin with radish, onion, lentils, garlic, beans, and bean green manure, did not have a significant effect on cumin wilt (33).

Using bacillus subtilis as an antagonist bacteria separated as a Fusarium wilting agent from contaminated soil, inhibit the growth of *F. oxysporum* wilting of cumin in agar culture medium (30).

Sodium Methyl Dithiocarbamate (SMD) has a positive effect on seed germination, plant growth and prevention of cumin wilting. Increasing the concentration of SMD solution up to 50%, 20 days prior to planting, resulted in greater seed germination and better establishment of young seedlings. It also enhances plant growth and reduces fusarium wilting infection (33, 29). Investigation of the role of seed contamination, caused by the spreading of the disease agent, indicates that due to seed contact with infected soil at the time of harvest, the fungus is transferred from the disease agent to new areas (31). The possibility of disease transmission via seeds shows that the agent is soil borne and located within the endosperm (48).

Evaluation of different fungicides in sterilized seeds in India before planting and study of the effect of fusarium wilting on growth, yield, and tolerance of cumin indicated that agrosan, which is a soil persisting fungicide, increases yield of cumin significantly. Other fungicides such as ceresan, benomyl and thyram, even when seeds were highly contaminated with mycoflora, resulted in higher yield (47).

The reaction of different varieties of cumin against three different races of the disease agent indicates that all understudied cultivars were susceptible to the disease agent (49). However, Champawat (5) in two years of field study on 161 varieties of cumin found four which were relatively tolerant, three were moderately susceptible, and the rest of them were extremely susceptible to wilting.

Use of organic manure, nitrogen, Phosphorus, and potash fertilizer reduced the intensity of disease, and use of 30 kg N2 ha^{-1} and 30 kg k$_2$o ha^{-1} decreased fungal infection in a study by Champawat et al. (6, 22). The least amount of contamination and highest yield were obtained by using 30 kg nitrogen, 20 kg phosphorus and 30 kg potash per ha (6). Two years of cultural practical experiments to control this disease showed that summer tillage has a significant effect in controlling this disease. By increasing the number of tillage, the severity of disease reduced (34, 35, 40). Using different soil chemicals to control cumin wilt in two years of field study showed that carbophoran, phorite, and tamik were very effective in controlling cumin wilting disease (7).

In Iran, the first attempt to understand the type of cumin wilting agent, the different factors affecting it and how to control the disease, was made by Alavi (1). He announced the disease agent is *F. oxysporum* f. sp. cumini Prasad & Patel. In a series of experiments, he investigated the effect of irrigation, fertilizer, seed treatment, planting date and amount of seed in Sabzevar and neighboring cities in Khorasan; the result of his studies indicated that delaying planting date from December to March reduces the disease damage to cumin. The amount of seeds/unit area had no effect on disease reduction. None of N$_2$, P$_2$O$_5$, and K$_2$O fertilizer reduced the disease. Seed treatments with understudied fungicide also did not control the disease. To investigate the wilting disease, Hagian (16) conducted two years' study in Khorasan area; he took fungus samples at different stages of growth mainly prior to flowering. The fungi were detached from roots and stem segments of contaminated plants and their pathogenicity was studied using two culturing methods, culturing on silt and culturing on cornflour. He applied the spore suspension to sterilized soil before planting, and at the basal part of healthy plants. The results indicated the pathogenicity of only two species of *fusarium*-like fungi. The species *F. oxyosporum* f. sp. cumini Prasad &

Patel formerly has been reported from India as cumin wilting disease (37), and the species of *F. solani* (Mart.) Appel. Wr. was another wilting of cumin agent reported from Iran (20). The severity of isolation of *F. oxysporum* varied from 20% to 620%. Isolation of fungus disease agent is able to move within xylem tissue systems and causes infection in different plant organs. The severity of wilting disease was measured in different parts of Khorasan region. The highest and lowest amount of infection was observed in Ferdows with an average amount of 28.7% and Torbat-e-Heydariyeh with 14.5% respectively. In this study the possibility of seedling death prior to emergence from the soil was evaluated and it was observed that *F. oxysporum* is able to attack the seedling within the soil and results in death of the seedling. The transmission of the disease agent via seeds was evaluated by studying superficial and internal tissues of seed. The presence of fungus hyphae inside the endosperm was approved. The results showed that those seeds with internal contamination did not germinate and they perished soon after planting (19).

6-1-1-1 *Diagnosis*

The primary symptom is wilting of the tip of growing leaves. The infected leaves usually turn yellow or sometimes desiccate without discoloration. In the early stages of disease, the root of the infected plant is similar to healthy plants but sometimes brown scars may be observed on the root surface. In the progressive stage of disease, the lateral root and root epidermis may get destroyed. Brownish discoloration appears on vascular bundles, and fungus mycelium, which is primarily restricted to the vascular system, may attack the cortex parenchyma in the progressive stage. The disease symptom appears about four to five weeks after planting, when the seedlings are in six-eight leaf stage and continuous throughout the plant growth. The plant is sensitive to disease agent at all times during growth stages. The disease looks like scattered patches of different sizes across the field, which may spread radially in all directions and result in severe damage to the plant and yield reduction (16).

6-1-1-2 Biology

The disease agent usually attacks the plant after winter frost mainly -
during late March. The spreading rate of the disease depends on
environmental conditions and host growth stage. If the infection
happens during early growth stage of the host when the plant is
extremely susceptible, the seedlings will perish in less than 10 days and
the disease may progress within the field and destroy the crop.
However, in the case of a long winter, the beginning of the disease
agent outbreak coincides with vigorous cumin growth and, to a large
extent, it is able to withstand the infection. The agent is soil borne
and is transmitted easily via seeds. The disease agent can spend
winter within the soil as chlamydospore, or spore and hyphae on the
surface or within endosperm respectively (16, 23, 38).

6-1-1-3 Control

Since this is a soil borne disease, the control measure of the disease is
based on prevention. The following suggestions are recommended to
reduce the damage of disease infection:

1. To meet the challenge in order to improve and introduce
   cumin resistance to *Fusarium* wilt would be an effective
   method of disease control.
2. Use proper crop rotation and avoid planting cumin every year.
3. After harvesting the crop, all plant residues should be collected
   and burned to reduce disease infection in the subsequent year.
4. Since the disease is patchy in the early stages, the infected
   plants should be collected and burned to eliminate the
   spreading of spots during the current year and following year.
5. Do not use agricultural machinery already used in infected
   fields unless the wheels have been washed and sterilized with
   lime.
6. Since the disease is transmitted via seeds, it is better to sterilize
   the seeds with systemic chemicals such as benomyl.
   Carbendazim and Rural may be used against probable infection
   and transmission to uncontaminated fields.

6-1-2　Cumin Blight (*Alternaria burnsii*)

Cumin burn was reported for the first time from India (50), followed by reports from other countries (2). The agent known as cumin blight or *Alternaria burnsii* was reported from almost all countries which planted cumin in high acreage (36). Precise information is not available about the history of this disease in Iran, but with increasing cumin acreage since 1978, *Alternaria* blight of cumin spread drastically. So far, efforts to identify resistant types against the disease were futile. Several studies are being carried out to prepare effective chemicals or to evaluate chemicals against the disease by investigators. Therefore, considering the continuation of cumin planting, it is necessary to meet the challenge for disease control. The following researches concerning cumin burn are carried out throughout the world. Patel (37) examined the effect of Atrazine, thiobencarb, diuron, pendimethalin, 2,4,-D ester, fleuchloralin, metribuzine, and trifluralin with 5, l, and 1.5 times the usual dosage against cumin blight in vitro condition and concluded that the maximum inhibition of mycelium growth (79.8%) was obtained by using 1.5 times the normal metribuzine. In another experiment, Patil (41) examined the effect of plant age and seeding time against spreading of the disease. He concluded that in Anand, India the severity of disease in November was more compared to December or January, the least severity of disease was obtained in the second week of December.

Baswana (3) studied five planting dates (5 November , 25 December, 15, 25 and 4 January) and three fungicide (mancozeb, captafol, and carbendazim) on quality, yield, and blight infection of cumin. The results indicated that in the last planting date (4 January) disease infection was not present, the seed quality was better, but the yield was less than other treatments. In early planting when the disease severity varied between 55% and 96%, and captafol and mancozeb reduced the disease only during one or two years when they were used, carbendazim did not have any effect on disease control. In another research, Gemawat (11) studied the effect of different fungicides on disease control efficacy and concluded that organic fungicides are more effective than minerals or antibiotics. He reported that every other week spraying ziram (80%) with the ratio of 1/1,000, and zineb (65%) with the ratio of 2/1,000 controlled the disease by 72.6% and 61.7% respectively. Sankhla and Mathur (46) studied the effect of 11 fungicides under glasshouse conditions and concluded that except copper oxychloride, cupravit, and phytolan, the rest of the fungicides could control the disease equally effectively as other efficacious fungicides. However, ziram (80%) and captan (78.5%) with the ratio of 1.5/1,000 were better than other treatments.

The combined effect of sterilizing seeds with fungicide before planting and postplanting spray with benomyl, carbendazim, carbendazim-iprodion, etrimol, zineb, Bordeaux mixture, and Macoprex 10 days and 40 days after planting confirmed that all the fungicides were also used as seed sterilants and increased the yield to acceptable levels. Among the other fungicides, carbendazim was more effective in disease control, with the exception of macoprex and carbendazim- iprodion (25), which were extremely effective.

To understand the factors that were responsible for spreading of the disease in fields, Gemavat (13) carried out a comprehensive study and showed that 90% relative humidity along with 23° to 28°C for three consecutive years resulted in spreading of the disease. The disease usually spread in the direction of the wind. Dust pollution and plant debris caused the persistence of disease within the field. The disease agent was transmitted via seeds and the number of spores released during early morning increased gradually. Gemawat (12) showed that cumin is sensitive to disease agent only after flowering. To establish this point, the amounts of carbohydrates and different amino acids were measured and it was found that this is a low sugar disease. Since the best sources of carbon for growth of disease agent in non-living environment are maltose and sucrose, and during blooming and expansion of disease the amount of these substances increase within the plant, they may probably have an important role to play in plant metabolism and sensitivity to cumin blight.

The first study to control the disease in Iran was conducted by Hagian and Jafarpour (17). They investigated the effect of nine fungicides: benomyl, dodin, captan, mancozeb, carboxin-thyram, iprodion- carbendazim, methyl thiophenat methyl, and copper oxychloride in Potato-Dextrose Agar (PDA) media with 10, and 100 ppm concentration to inhibit the mycelium growth of disease agent. They found that the following six fungicides, i.e. dodin, captan, mancozeb, carboxyn-thayram, iprodion-carbodazim, and carbendazim had the best effect in prohibiting mycelium growth at 100 ppm concentration with 75.9, 70.3, 44.4, 39.2, 35.5, and 11.8% respectively. They selected these fungicides and used them against the cumin blight disease in a field study. The results indicated that all fungicides were able to control leaf and stem contamination at $P<$ 5% level and resulted in increasing the yield of cumin seeds, however, significant differences were not observed between the time of spraying, fungicide concentration, and the interaction between them. They recommend using any one of the fungicides, i.e. mancozeb, captan, or Eprodion- carbendazim, at early stages of contamination to control the disease. During flowering, which is the most sensitive stage of the host to the disease, they recommend the use of one of the fungicides iprodion-carbendazim, captan, Mancozeb, and carboxyn- thyram, based on priority at three weeks interval with concentration of 2 grams per litre just after observing the first symptom of disease (18). Table 6-1 shows the reduction in contamination of leaves, branches and flowers and increasing yield due to using different fungicides.

Table 6-1 *Effect of different fungicides on percent reduction of leaves, stems and flowers, and increasing of yield (18)*

	Contamination reduction (%)		
Fungicide	*Flowers*	*Leaves and stems*	*Increasing yield (%)*
Captan	20.1	20.30	7.7
Carbendazim	21.2	22.95	64.7
Mancozeb	20.2	20.17	73.0
Carboxyn-Thyram	21.7	21.00	61.0
Dodin	22.7	21.50	44.5
Iprodion-carbendazim	11.7	16.70	70.0

6-1-2-1   *Diagnosis*

The disease symptom differs based on the time of contamination and appearance of disease. It begins with the appearance of necrotic spots and ends with the death of the plant. Usually the first necrotic spots start on the tips of the leaves and spread to the remaining part of the leaves. When the climatic conditions are favorable, the spots expand rapidly. The disease at the junction of branches to the main stem looks like black necrotic spots. During early spring when relative humidity is high and the weather is hot, the disease expands more rapidly and during flowering stage the lower parts of the plant turn black and the plant dies (17).

6-1-2-2   *Factors Affecting Epidemics*

The study of realizing the expansion and disease outbreak in cumin production areas shows:

1. The disease appears only after flowering and the plant is not sensitive to the disease before blooming. The sensitivity exists from flowering to harvest.
2. The disease progresses during cloudy and humid conditions, and if it rains after infection, the disease spreads rapidly. A shower after flowering results in suitable condition for infection.
3. Hot and dry condition is not favorable for disease and it is a means of disease control.
4. The disease is usually observed in new grown cumin areas, as well as areas where cumin has not been planted during the last 10 years. This is an indication of the possibility that the disease is seed borne.
5. The long period of planting (December to March) results in: (1) the host becoming easily accessible by disease agent, and (2) spreading the disease during favorable conditions.
6. In the fields in which cumin is planted successively for more than a year, the spreading of the disease increases the following year.

Mathur and Mathur (31) performed experiments relating to the transmission of disease agent via seeds and plant debris. They isolated the blight agent fungus disease from three samples of cumin field soils. The complementary studies on determination of the role of soil in the beginning of cumin blight as a primary source of inoculum by planting cumin within pots of sterile soil already infected with disease agents indicated that 40% of the plants were infected 10 days after flowering. One year later, the fungus disease agents which were collected from infected plant debris and kept under laboratory conditions, were isolated and placed in cumin pots with sterile soil and healthy seeds and plants. The disease symptom was observed 20 days after flowering. These studies indicate that pathogen remain alive either in soil or plant debris, and are responsible for infection. New areas or areas with no cumin records during the past 10 years also possessed the seed borne disease agent. Uppal also determined that seed infection plays an important role in constant fungus persistence and at the beginning of the primary infection (50). A study concerning seed mycoflora conducted by Gemawat (11) indicates that the disease agent not only infects the seed surface but also penetrates into the seeds, and the seed is contaminated externally and internally with the disease agent. The long period of planting in cumin plantation areas also results in cumin infection. The results obtained about the first appearance of disease and adjusting it with plantation date in different areas of cumin plantation shows that the infection occurs almost two months after plantation. These studies clearly indicate that long range of cumin planting results in spreading of the disease agent to adjacent areas where cumin is planted later. Humidity and precipitation causes the outbreak of disease under laboratory and field conditions. The disease spreads rapidly under conditions of 90% relative humidity and it will be more severe if the humidity continues for three days. By reducing the humidity and appearance of the first symptom of hot and dry condition, the disease spreading ceases (13).

The investigation by Uppal (50) indicated that A. *burnsii* can grow in a wide range of temperature from 5° to 35°C and the optimum temperature for cumin growth and sporolation is around 26° to 27°C. This fungus is able to stay alive for 12 to 18 months by storing between 5° and 8°C in a refrigerator. Such conditions are present in most cumin plantation areas for growth, development and persistence of disease agent. The hot and dry climate is not favorable for multiplication of the disease agent and prevents the spreading of the disease. It seems that under non-living conditions, a specific correlation exists between expansion of disease, temperatures and growth of pathogens (mycelium growth, sporolation, and spore germination). Such issues are also observed in other disease agents in which species of *Alternaria* are involved (45). Wind direction has an important role in disease spreading and the direction of disease infection. Usually the direction of disease spreading is aligned with the direction of dominant winds, resulting in the infection of disease to adjacent areas.

The determination of the number of *Alternaria* spore population in cumin field space during 00.00 to 24.00 hours shows a peak of spores between 09.00 and 12.00 hours in March and April, and the number of spores followed a specific pattern in the morning with daily alteration. The number of spores reduces gradually during noon and afternoon, with another increase at night, and a peak at 09.00 to 12.00 hours of the next day (13).

6-1-2-3   Control

The following methods are recommended to control A. *burnsii*:

1. Attempts should be made during breeding programs to breed and introduce cumin resistant variety to A. *burnsii*.
2. Since this too is a seed borne disease and also transmitted via seeds, sterilizing the seeds before planting by use of fungicides such as mancozeb and captan, at the rate of 1.5 gram per litre can be effective in disease control.
3. Spraying the plants as soon as the fungus infection is observed. The priorities are commercial chemicals of mancozeb, captan, or iprodion-carbendazim mixture during flowering when the

plants are most sensitive to infection, at three weeks interval at the rate of 2 grams per litre.

6-1-3   Powdery Mildew *(Erysiphe polygoni)*

This disease has not been reported from Iran yet, but has been reported from India with the rate of infestation between 50 and 100%. The symptoms of the disease are the appearance of white powdery spots on the surface of the leaves or stems. The disease agent also attacks the flower, which results in destroying or malformation of the fruits. The agent of the disease is known as *Erysiphe polygoni*. To control the disease, Mathur. (26) evaluated six different fungicides against this disease and concluded that elosal and caratan at the rates of 77.1 and 71.8% were the most effective in controlling powdery mildew.

6-2   PESTS

The most harmful pest reported from cumin is the green aphid of peach. Since this polyphagus pest is considered as one of the most destructive pest crops in Iran, the characteristics and control measures of this pest are described below (15).

6-2-1   Peach Green Aphid *(Myzus persicae)*

Peach green aphid is a polyphagus pest which attacks several fruit trees and ornamentals. This pest is active on leaves and young stems and spread throughout Iran. This aphid is one of the main carriers of plant viruses and is holocyclic with two hosts. During spring, the parents of wingless and viviparous aphids emerge from hatched eggs. In early summer, the winged viviparous females migrate from trees to their secondary host where winged and wingless generations are observed. During fall, the winged aphids reappear and migrate towards peaches and other trees where male and female aphids appear, and female aphids after fertilization, lay eggs on trees.

Investigation of the damage caused by this aphid on cumin showed that most of the damage occurs via aphid populations that attack cumin during flowering, and the damage is less during seed setting. Therefore, the best time to control this pest is at the blooming stage (15).

This aphid is easily controlled by using systemic pesticides such as metasystox, or dimethoate, or contact pesticides such as diazinon, malathion, or gusathion (4, 43).

6-2-2   Ants

Ants are another pest that attack cumin at a late stage of growth when the seeds are collected easily from umbels. Ants occasionally and abruptly attack cumin fields, and collect most of the seeds within a few days in such a way that nothing remains in umbels except a chaff, and the white color of umbels is an indication that the seeds have been carried away by ants.

6-3   WEEDS AND WEED CONTROL

Cumin is a poor competitor against weeds and if weeds are not managed and controlled, the yield of cumin is reduced drastically. Unfortunately not much study has been carried out on weed control in cumin. Gora (14) in a study in India, concluded that lambsquarter (*Chenopodium album*), *Chenopodium murale*, and bermuda grass (*Cynodon dactylon*) are the most important weeds in cumin. Delghandi (9) identified that the most harmful weeds in cumin fields of Khorasan (Iran) could be ranked in the following order: Santalin yarrow (*Achillea wilhelmsii*), hoary cress (*Cardaria draba*), lambsquarter (*Chenopodium album*), field bindweed (*Convolvulus arvensis*), barnyard grass (*Echinochloa crusgalii*), knot weed (*Polygonum aviculare*), green foxtail (*Setatia viridis*), and wild mustard (*Sinapis arvensis*).

It seems that due to the type of growth and low Foliage Area Index (FAI) of cumin in the initial stages, the weeds should be controlled in the early stages to obtain higher yield. Kafi (24) concluded that weed control three weeks after cumin emergence is acceptable. He observed little difference in yield between one weeding compared to two or more weedings but statistically it was not significant. When cumin is three weeks old, weeds are mostly around 5 cm and it is a good time for weeding and thinning of cumin (41). Hosseini (21) in a study on the critical period of weed control, concluded that weeding cumin 24 to 38 days after emergence did not decrease the yield. She also concluded that weeding during early stages of growth was preferred to controlling weeds at later stages of growth.

Abstracts of studies carried out on the competition of weeds with cumin have been published in the annual conference of the weed identification society in India, according to which the critical period for cumin weed control has been reported as 15 to 30 days after emergence (23).

Some studies concluded that one weeding has a better effect than weedy check, but if this weeding is repeated once more, the result will be even more favorable (Table 6-2).

Application of herbicides for weed control is promising. Few studies indicate that chemical control does not differ from hand weeding (42). Chaudhary in India concluded that hand weeding of cumin yielded 336 kg ha^{-1} and application of the herbicide terbutryne resulted in 322 kg ha^{-1}, whereas in the control only 47.2 kg ha^{-1} cumin was obtained (8).

Table 6-2 *Biological yield, seed yield, straw yield, and harvest index of different weed control treatments of cumin (43)*

Treatment	Biological yield (Kg/ha)	Seed yield (Kg/ha)	Straw yield (Kg/ha)	Harvest index (%)
Weedy check	728	240	388	46.6
One hand weeding	1,281	683	597	53.4
Two hand weeding	1,578	809	769	51.3
Three hand weeding	1,613	832	781	51.6
LSD 0.05	323	162	162	1.4

Rathore (44) in the project of weed control measures on growth and yield of cumin, suggested that pendimethalin and basalin at 1 kg ha^{-1} preplant pre-emergence and soil incorporated; fleuchloralin at 0.5 kg ha^{-1} preplant and soil incorporated, are among the best treatments for weed control in cumin.

Another experiment was conducted at Gujarat University (India) to study the effect of chemicals on weed control. The results showed that hand weeding, and oxadiazon at 1 kg/ha and pre-emergence had higher yields. Treatments of fleuchloralin (preplant and soil incorporated), oxyfleurophen (pre-emergence) and benthiocarb (pre-emergence) also had a good effect (23). Rahimi (42) examined the spatial and temporal effect of 30 different herbicides on cumin and recommended the following cases:

1. Pre-plant soil incorporated includes ethalfluralin, trifluralin, dinitramin, pendimetalin and EPTC. He recommended that sowing preferably be done 10 days after herbicide application for optimum growth of cumin.
2. Pre-emergence herbicides include chlorbromuron, metachlor, prometryn, chlorethaldimethyl, simizin, and oxadiazon.
3. Post-emergence herbicides include chlorbromuron, Prometryn, and linuron. Using isoproturon at rates of 1 and 2 kg/ha and oxyfleurophen at rates of 0.125 and 0.25 kg/ha preplant or pre-emergence showed higher yield similar to two times hand weeding (23).

In a study during 1981 to 1986 relating to "the effect of weed control on growth and yield of cumin" using pre-emergence and soil incorporated herbicides of pendimethalin and also fleuchloralin at rates of 1 kg/ha resulted in effective control of bermuda grass, heliotrope, and lambsquarter.

In another study during 1987, Indian scientists concluded that preplant soil incorporated with oxadiazon at rates of 1 kg/ha in combination with fleuchloralin at 0.9 kg/ha and oxyfleurophen (0.48 kg/ha) with benthiocarb (2 kg/ha), had significant differences with weedy check and yield was also satisfactory.

Cumin is not cultivated in Europe and the United States, and adequate data regarding different aspects of cumin from these continents is not available. However, for weed control in dill, which is a close relative of cumin, application of pre- emergence or preplant prometryn (1.8 kg/ha) when weeds are less than 5 cm tall is recommended. For caraway, a very close relative of cumin, the following herbicides are recommended (as per Guide to Crop Protection 2000, prepared by Manitoba Agriculture and Food):

1. Preplant soil incorporated with ethalfluralin 60%, commercially known as Edge – Dc, at the rate of 1.4 kg/ha for soils with 2 to 4% organic matter, and at rate of 2.1 kg ha^{-1} for soils with 4 to 15% organic matter and soil volume 120 lit ha^{-1}.

2. Preplant soil incorporated spring application of ethalfluralin 5%, commercially known as Edge Granular, at rates of 17 kg ha^{-1} for soils with 2 to 7% organic matter, and 25 kg ha^{-1} for soils with 4 to 15% organic matter.

3. Post-emergence application of linuron, commercially known as Afalon, (1.26 to 1.65 kg ha^{-1}) during two to four leaf stage. Irrigation or precipitation seven to ten days after application of herbicide will be useful. Using Afalon in combination with CPA is not recommended.

4. Using sethoxydim (450 g/litre) commercially known as Post-Ultra, at rates of 0.65 lit ha^{-1} in 100 to 150 litres of water.

Note that these recommendations are for caraway, a very close relative of cumin. It is recommended that for cumin preliminary experiments be conducted for a precise conclusion.

Experiments with preplant application of trifluralin 33% (treflan) and ethalflvralin 45% (sonalan) are underway and the results are promising. However, they are not yet in the state of recommendation.

6-4 SUMMARY

Cumin is very sensitive to fungal diseases. The disease agent of cumin wilting is *Fusarium oxysporum,* a common disease in Iran and India. The leaves of infected plants turn yellow and the vascular system turns brown. The disease agent is seed borne and is transmitted easily via seeds. The cumin wilting disease can overwinter as chlamydospore or hyphae. A control measure against this disease is based on prevention. Practices such as introduction of resistant varieties, crop rotation, burning plant residues, removing and burning the infected plants, and sterilizing seeds with systemic fungicides, reduce damage to the crops. *Alternaria* is very common in most cumin production areas. It starts by production of necrotic spots on the foliage and ends with the death of the plants; expansion being at the peak during cloudy days of spring with relatively high humidity. Cumin is sensitive to this fungus after flowering until maturity. Pathogen stays alive in soil and is responsible for reinfecting the host. The disease agent penetrates from the surface into the seeds. Using resistant varieties, sterilizing seeds with the appropriate fungicide, and spraying the infected plants with fungicide, will help in reduction of disease injury. There are extensive challenges in identifying the best herbicides for cumin weed control, but due to different geographical conditions in planting areas, type of soil and diversity of weeds, a common and broad-spectrum herbicide is not introduced. Further researches relating to the effect of herbicides on cumin are underway. On the whole, it may be concluded that:

1. Dinitroanilin herbicides such as ethalfluralin, trifluralin and basalin fleuchloralin, are probably effective if uses preplant and soil incorporated.

2. Herbicides such as oxadiazon, prometryn, simazin, pendimethalin, isoproturon. oxyfleurophen and benthiocarb, preplant or pre-emergence probably produce the best results.

3. Herbicides such as linuron, and Acetyl Co enzyme A (ACCase) inhibitors such as sethoxydim, may be useful. However, in all cases, particularly ACCase, resistance to herbicides should be observed.

REFERENCES

1. Alavi, A. 1969. Fusarium disease of cumin. Plant Pests and Disease J. Tehran, Iran. vol. 5(3): 92-98.

2. Bandopadhyay, B. 1980. Incidence of a blight disease of cumin in west Bengal.Science and Culture. 46(9): 341- 342.

3. Baswana, K. S. 1991. Effect of sowing date and fungicide on seed quality, yield and disease incidence of cumin. Indian Cocoa Arecanut and Spices Journal. 14(4): 155- 157.

4. Behdad, A. 1992. Iran crop pests. 3 rd . Neshat Publication, Esphahan, Iran.

5. Champawat, R.S. 1990. Field screening of cumin germplasm against *Fusarium oxysporum* f.sp. cumini. Indian Cocoa Arecanut and Spices Journal. 13: 142.

6. Champawat, R.S. and V.N. Pathak. 1988. Role of nitrogen, Phosphorus and potassium fertilizers and organic amendmends in cumin (*Cuminum cyminum*) wilt incited by *Fusarium oxysporum* f.sp. cumini.Indian J. Agric . Sci. 58(9): 728- 730.

7. Champawat, R. S. and V. N. Pathak. 1988. Soil application of different insecticides and nematicides for the control of wilt of cumin. Indian J. Plant Prot. 16: 195- 196.

8. Chaudhary, G. R. 1989. Effect of nitrogen level and weed control on weed competition, nutrient uptake and quality of Cumin. Ind. J. Agric. Sci. 59: (6):397- 399.

9. Delghandi, M. 2004. The weed flora of cumin fields. Proceedings of the 1st Cumin National Symposium in Sabzevar, Iran. pp. 99-100.

10. Gaur, M. M. 1949. Diseases of cumin and fennel. Plant protection Bull. 1: 20- 21.

11. Gemawat, P.D. 1969. Efficacy of different fungicide for the control of Alternaria blight of *Cuminum cyminum*. Indian Phytopathol. 22(1): 49- 52.

12. Gemawat, P. D. 1971. Alternaria blight of *Cuminum cyminum* physiology of pathogenesis. Indian Phytopathol. 38: 38-43.

13. Gemawat, P. D. 1971. Epidemiological studies on Alternatia blight of *Cuminum cyminum*. Indian Journal of Mycology and Plant Pathology. 2(1):65-75.

14. Gora, D. R., N. L. Meena, D. L. Shivran and D. R. Shivran. 1996. Dry matter accumulation and nitrogen uptake in cumin (*Cuminum cyminum*) as affected by weed control and time of nitrogen application. Ind. J. Agron. 41: 666-667.

15. Gupta, B.M. and C. P. S. Yadava. 1989. Incidence of aphid *Myzus persicae* on cumin in relation to sowing date variation. Indian J. Entomol 51(1): 60-63.

16. Hagian, M. 1994. The study of cumin wilting in Khorasan. M. Sc. Thesis in crop protection, Tehran University, Karaj College of Agriculture, Karaj, Iran.

17. Hagian, M. and B. Jafarpour. 1996. Chemical control against *Alternaria burnsii*, Final Report. Organization of Technology and Science, Mashhad, Iran.

18. Hagian, M. and B. Jafar pour. 1998. The study of fungicide effects on cumin blight. Proceeding of 13th. Plant Protection Congress, 2-7 Sep. 1998. Karaj, Iran.

19. Hagian, M., G, Zad , G. Hajaroud, A. Sharifi Tehrani and M. Rastegar. 1995. The study of the method of cumin disease transplanting. Proceedings to 12th Plant Protection Congress. 2-7 Sep. 1995. Karaj, Iran.

20. Hagian, M., G.Zad, G. Hejaroud and A. Sharifi Tehrani. 1994. Introducing the cumin damping off disease agents in Khorasan areas. 2nd Saffron and Medicinal Plant Gathering, Gonabad, Iran.

21. Hosseini, A. 2005. Determination of critical period of weed control in cumin medicinal plant. M. Sc. Thesis, Ferdowsi University of Mashhad, College of Agriculture. Mashhad, Iran.

22. Howard, R. J. and S. F. Blade. 2000. Crop diversification centers. 2000 annual reports. Crop Diversification Center, Alberta, Canada. In http://www. Agric. gov. ab. ca.

23. Indian Society of Weed Science. 1985. Chemical control of cumin. Abstract of papers.

24. Kafi, M. and M. H. Rashed-Mohassel. 2000. The effect of population density and times of weed control on growth and yield of cumin. Journal of Science and Technology. Mashhad, Iran. Vol 6. 151- 158.

25. Lakhtaria, R. P. 1979. Evaluation of fungicides against blight of cumin. Indian Phytopathol. 44.

26. Mathur, R. L. 1971. Evaluation of fungicides against powdery mildew disease of cumin caused by *Erysiphe polygoni*. Indian Phytopathol. 24(4): 796- 798.

27. Mathur, R. L. and B. L. Mathur. 1965. Annual report of scheme for research in wilt disease of Zeera in Rajasthan. Department of Agriculture, Rajasthan.

28. Mathur, B. L., and R. I. Mathur. 1966. Metabolites of *Fusarium oxysporum* f.sp. cumini in relation to cumin wilt influence of growth period , pH and dilution. Indian Phytopathol. Vol xx 42-44.

29. Mathur, R. L., and B. L. Mathur. 1967. Effect of SMDC on germination plant growth and Fusarium wilt incidence of cumin. Plant Disease Rep.51 (8): 629- 631.

30. Mathur, B. L. and R. L. Mathur. 1968. Occurence of bacteria antagonistic to *Fusarium oxysporum* f. sp. cumini and other soil fungi in cumin wilt sick soil. Prot. Nat. Acad. Sci. India. 38(B): 49-52.

31. Mathur, B. L. and R. L. Mathur. 1970. Role of contaminated seeds in dissemination of cumin wilts fungus *Fusarium oxyporum* f.sp.cumini. Raj. J. Agric. Sci.1(2): 79-82.

32. Mathur. B. L. and N. Prasad. 1963. Variation in *Fusarium oxysporum* f.sp. cumini in nature. Indian J. Agric .Sci. 34:273-277.

33. Mathur, B. L. and N. Prasad. 1964. Studies on wilt disease of cumin by *Fusarium oxysporum* f.sp. cumini. Indian J. Agric. Sci. 34:131-137.

34. Mathur, B. L, H.C. Sankhala and R. L. Mathur, 1967. Influence of cultural practices on cumin wilt incidence. Indian Phytopathol. Vol xx. 32-34.

35. Paihar, G. N. and R. Singh. 1994. Effect of cultural and herbicidal weed management on the yield of cumin. Annals of Arid Zone. 33(4): 309-312.

36. Patel, R. M. 1971. Alternaria blight of *Cuminum cyminum* and its control. India Phytopathol. 24(1): 16-22.

37. Patel, R. J. 1993. Evaluation of herbicidal concentrations against *Alternaria sp*. causing cumin blight. Proceeding of an Indian Society of weed science International Symposium. 18-20 November. vol. II 143-144.

38. Patel, P.N. and R.L. Prasad. 1957. Fusarium wilt of Cumin. Curr. Sci.26:181-182.

39. Patel, P.N. and N. Prasad. 1963. Fusarium wilts of Cumin (*Cuminum cyminum*) in Gujarat state, India Plant Disease Rep. 47:528-531.

40. Pathak, V.N. 1990. Management of cumin wilts by summer ploughing. Indian Cocoa, Areca Nut and Spices Journal. 13:107-108.

41. Patil, R.K. 1983. Age of the crop and the sowing period on the incidence of cumin blight. Indian Journal of Mycology and Plant Pathology. 13(3): 107-108.

42. Rahimi, M. 1993. Chemical control against weeds in cumin. Organization of Technology and Science, Mashhad, Iran.

43. Rastifard, A. 1971. The study of legumes aphids in karaj. M. Sc. Thesis crop protection, College of Agriculture University of Tehran, Karaj, Iran.

44. Rathore, P. S. 1990. Effect of weed control measures on growth and yield of cumin. Indian J. of Agronomy.

45. Rotem, J. 1964. The effect of weather on dispersal of Alternaria spores in a semi-arid region of Israel. Phytopathology, 54:628-632.

46. Sankhla, B. A. and R. R. Mathur. 1973. Evaluation of fungicides against *Alternaria* blight disease of cumin. Indian Phythopathol. 26(1); 154-155.

47. Singh, R. D. 1978. Evaluation of seed dressing fugicides for their effect on the stand, growth and yield of cumin in field, Indian Phythopathol. 30:198-201.

48. Singh, R. D., L. Chaudhary and G. Kanaiyalal 1972. Seed transmition and control of Fusarium wilt of cumin. Phythopathol. Medit. 11:19-24.

49. Sharma, A. K. Sharma. 1982. Reaction of varieties of cumin against *Fusarium oxysporum* f.sp.cumini, Indian Cocoa, Arecanut and Spices Journal 8 .

50. Uppal, B. N. and M. K. Patel. 1938. Althernaria blight of Cumin. Indian J. Agri. Sci. 8: 49-62.

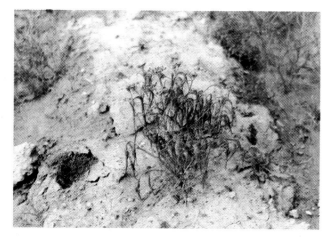

Photo 6-1 Cumin wilt

Photo 6-2 Cumin leaf blight

Photo 6-3 Cumin leaf blight

CHAPTER 7

Genetics, Breeding and *in vitro* Production of Cumin

A. Bagheri[1] and A. Mahmodi[2]
1 abagheri@ferdowsi.um.ac.ir
Faculty of Agriculture, Ferdowsi University of Mashhad, Iran
[2] aampira@yahoo.com
Dryland Research Center of North Khorasan, Iran

7-1 KARYOLOGY OF CUMIN

Studies on cumin karyotype have shown that the chromosome number of cumin is $2n=14$. Baijal and Kaul (3) have also reported cumin chromosomes as $2n=14$ and stated that these chromosomes could be divided into seven morphological groups based on their absolute length and initial and secondary cavities. Cumin also has three pairs of satellite chromosomes. Jha and Roy (15) using tissue culture studies on cumin, showed that callus as well as regenerated shoots, roots and buds are diploid with 14 chromosomes. Cytological analysis of three cumin varieties by Chattopadhyay and Sharma (6) also showed that the number of somatic chromosomes is $2n=14$. Most chromosomes had cavities at their end or central parts.

Despite evolution of different cumin varieties, karotypic analysis has shown basic morphological similarities with no distinct variation in length or volume of chromosomes between varieties. In addition, study of DNA content showed stability of genotypes with little variability in different varieties. However, DNA content increased in proportion to length or volume of chromosomes (6).

7-2   GENETIC DIVERSITY

Quantitative analysis of genetic variation in 50 cumin varieties (4) for yield and five growth characters showed that yield has the highest coefficient of genotypic and phenotypic variation (23.41 and 24.17%, respectively) and heritability (93.82%). El-Ballal (10) studied the coefficient of variation of eight yield components, correlation between yield components and genetic improvement in yield components of cumin in two locations. He also studied the effects of 100 ppm indol acetic acid (IAA) on cumin fertility when plants were treated at rosette stage. The results showed that 75.41% of 40 treated individuals were fertile and the expected genetic improvement was 95.95%. Avatar et al. (2) studied 13 morphological traits of 27 Indian cumin lines as well as lines from Egypt and Libya. Genotypes were clustered based on their genetic variability. Clusters I and III with respectively, 24 and 3 genotypes, all from India, showed the highest similarity. However, genotypes from Egypt and Libya were grouped in different clusters. It was hypothesized that genetic and geographical variations are correlated. However, there are exceptions to this hypothesis. Cumin yield and days to maturity had respectively, the highest contribution in the observed variation.

Dhayal et al. (9) studied variation in response of cumin germplasm to salinity in salt affected and normal soils. High value of genotypic and phenotypic variances, heritability and genetic improvement, number of seeds per umbrella and individual plant weight in salt affected soils indicate its additive effect on expression of traits.

7-3   EFFECTS OF IRRADIATION

In the study of Koli et al. (17) cumin seeds were subjected to gamma radiation levels of 20, 30, 40, 50 and 60 krad before planting. M2 generation was produced by self-pollination of M1, and 11 traits including yield were recorded. Some M2 lines showed improvements in number of days to flowering, days to maturity and individual plant weight. In another study (18) dry seeds of cumin (var. RZ10) were treated with 20, 30, 40, 50 and 60 krad gamma radiation using ^{60}Co. M1 generation was planted in field conditions and M2 seeds were harvested after self-pollination. For each treatment 10 elite individuals were selected and their yield and yield components were compared with control. Mean plant height had increased in all irradiation levels. However, its variance decreased (except in 30 krad treatment). Mean and variance of number of umbrellas and number of seeds per umbrella increased linearly with irradiation levels but seed yield and 1,000-seed weight decreased.

To achieve high yielding cumin cultivars, Khavari and Bagheri (16) in a five-year experiment selected 484 individuals from 147 landraces of Khorasan province in Iran. Statistical analysis on seed yield during successive years showed highly significant differences between selected lines, with the highest observed yield of 808 kg ha^{-1}.

7-4 *IN VITRO* CULTURE

Cumin production is restricted by several biotic stresses, among them *Fusarium oxysporum* and *Alternaria bursnsii* are of great importance (1, 19). There is limited potential genetic variation within cumin germplasm for classical breeding methods in particular for environmental stresses (5). Recent progresses in gene manipulation, DNA technology and gene transfer are promising methods for improvement of cumin resistance to biotic and abiotic stresses. Therefore, there is an increasing interest towards *in vitro* regeneration of cumin and further genetic manipulation of regenerated plants (7, 12, 22, 25).

Genetic transformation using *Agrobactrium* as a powerful tool for plant improvement could be used for producing transgenic cumin plants with high wilting resistance. Regeneration of plants from such explants during a short period of time and with minimal use of growth regulators will decrease the possibility of formation of somaclonal variants (23). `

7-5 COMPOSITION OF GROWTH MEDIA

Growth and regeneration of cumin explants in solid and semi-solid MS media, as well as in MSN solution culture, is reported (13, 16, 22, 24, 25, Table 7-1). In some researches, growth of explants improved when some minerals were reduced to half. Growth regulators such as NOA, 2,4-D, IBA, BA, IAA, TDZ, ABA, NAA, Kinetine and Adenine have been reported for *in vitro* cultures of cumin (13, 16, 21, 25). Rooting of regenerated cumin shoots would be possible in ½ MS media supplemented with 0.5 mg l-1 IBA and NAA (16).

7-6 SOURCE OF EXPLANTS FOR *IN VITRO* CULTURE

The main source of cumin explants is usually hypocotyl, shoot internodes, leaves and cotyledons. Tawfik and Noga (24) in a comparison of explants from hypocotyl and internodes, found that the best results for shoot regeneration were obtained from internodal explants. Tawfik and Gianinizzi (25) were able to regenerate cumin seedlings from callus formed from cotyledon and hypocotyl explants. However, other results showed better performance to hypocotyl compared to cotyledon explants (21). In comparison to explants from cumin hypocotyls and leaves, callus of hypocotyls had better growth (16).

Table 7-1 *Summary of in vitro studies on cumin*

Source of explant	Base media	Growth regulator	Additives	Growth response	Reference
Hypocotyl	MS	BA	PEG	Shoot regeneration	24
	MS	Kinetine, BA	-	Callus	22
	MS	BA, NAA, IAA, TDZ	-	Shoot regeneration	11
Hypocotyl, leaf	RS	2,4-D, NAA, BA, Kinetine	-	Flowering shoot	16
Cotyledon, Hypocotyl	MS	2,4-D, Kinetine	-	Seedling	25
Cotyledon	MS	Kinetine, IAA	-	Embryotic callus	22
Shoot	MS	BA	-	Shoot regeneration	24
Internodes	MS	IBA	PEG 6000	Rooting	24

7-7 SUMMARY

Cumin is a diploid species with $2n=14$. Yield and some yield components of cumin, e.g. days to maturity, show high variability. While gamma irradiation of seeds was not effective on yield improvement, there are promising results from *in vitro* culture of cumin explants. The best results were obtained when cumin shoots regenerated from hypocotyl explants. Explants from cotyledons and leaves also had good growth. Research on genetics and improvement of cumin is scarce. However, adaptation to a wide range of environments, increasing resistance of genotypes to abiotic (cold, drought and salinity) and biotic stresses (in particular fungal diseases) calls for more studies on genetics and improvement of this underutilized species.

REFERENCES

1. Agrawal, S., 1996. Volatile oil constituents and wilt resistance in Cumin (*Cuminum cyminum* L.). Cuur. Sci. 71: 177-178.
2. Avatar, R., S.I. Dashora, R.K. Sharma and M.M. Sharma, 1991. Analysis of genetic divergence in cumin (*Cuminum cyminum* L.). Indian Journal of Genetic and Plant Breeding 51: 280-291.
3. Baijal, S.K. and B.K. Kaul, 1973. Karyomorphological studies in *Coriandrum sativum* L. and *Cuminum cyminum* L. Cytologia 38: 211-217.
4. Baswana, K.S., M.L. Pandita and Y.S. Malik, 1983. Genetic variability studies in cumin (*Cuminum cyminum* L.). Haryana Agricultural University Journal 13: 596-598.
5. Champawat, R.S. and V.V. Pathak, 1990. Field screening of cumin germplasm against *Fusarium oxysporum* sp. Cuminin. Journal of Arecanut and Spices 13: 142.
6. Chattopadhyay, D. and A.K. Sharma, 1990. Chromosome studies and estimation of nuclear DNA in different varieties of *Cuminum cyminum* L. and *Carum copticum* Benth and Hook. Cytologia 55: 631-637.
7. Dave, A., A. Batra and R. Sharma, 1996. Origin and development of embryos produced from somatic tissues of cumin. Journal of Phytological Research 9: 65-66.
8. Hedge, C. and J.M. Lamond, 1972. Microsciadium Boiss. In: Flora of

Turkey-4. (ed. P.H. Davis) Edinburgh University Press, Edinburgh, UK., pp. 420-421.

9. Dhayal, L.S., S.C. Bhargava and S.C. Mahala, 1999. Studies on variability in cumin (*Cuminum cyminum* L.). on normal and saline soil. Journal of Spices and Aromatic Crops 8: 197-199.

10. El-Ballal, A.S.I., 1987. Cryptic polymorphism of sex expression in cumin (*Cuminum cyminum* L.). Acta Horticulturea, 20: 197-207.

11. Gupta, D. and S. Bhargava, 2001. Thidiazuron induced regeneration in *Cuminum cyminum* L. Journal of Plant Biochemistry and Biotechnology 10: 61-62.

12. Hussein, M.A., B. Amla and A. Batra, 1998. *In vitro* embryogenesis of cumin hypocotyl segments. Advances in Science 11: 125-127.

13. Jain, S.C., M. Purohit and R. Jain, 1992. Pharmacological evaluation of *Cuminum cyminum*..Fitoterapia 63: 291-294.

14. Jha, T.B., S.C. Roy and G.C. Mitra, 1982. *In vitro* culture of (*Cuminum cyminum* L.): regeneration of flowering shoots from calli of hypocotyls and leaf explants. Plant Cell, Tissue and Organ Culture 2: 11-14.

15. Jha, T.B. and S.C. Roy, 1983. Morphogenesis and chromosomal analysis in *Cuminum cyminum* L. Journal of Indian Botanical Society 62: 181-184.

16. Khavari K.S. and A. Bagheri, 2001. Evaluation of Iranian cumin landraces and their purification to obtain high yielding cultivars. Research Report, Ferdowsi University of Mashhad and Research Organization of Ministry of Jehad Keshavarzi, Iran, 40 p.

17. Koli, N.R., S.L. Dashora and E.V.D. Sastry, 2000. Genetic variability induced by gamma irradiation in M1 and M2 Generations of cumin (*Cuminum cyminum* L.). Indian Journal of Agricultural Science 70: 418-419.

18. Koli, N.R., S.L. Dashora, R.K. Sharma and D. Singh, 2000. Radiation induced variation in M2 generation of cumin. Annals of Agricultural Research 21: 448-449.

19. Omar, E.A., M.A. Nofal, S.M. Lashin and W.M.E. Haggag, 1997. Effects of some growth regulators on growth parameters and oil content of cumin with disease induced under two types of soil. Egyptian Journal of Horticulture 24: 29-41.

20. Rechinger, K.H. 1987. *Cuminum cyminum* L. In: "Flora Iranica" (ed. K.H. Rechinger) Akademische Druch-u. Verfaganstalt. Graz, Austria, pp. 140-142.

21. Shukla, M.R., N. Subhash, D.R. Patel, S.A. Patel and S.J. Eapen, 1997. *In vitro* studies in cumin (*Cuminum cyminum* L.). In: Biotechnology of Spices, Medicinal and Aromatic Plants. (eds. S. Edison, K.V. Ramana, B. Sasikumar and L.N. Babu). Proceedings of the National Seminar on Biotechnology of Spices, Medicinal and Aromatic Plants, 24-25 April 1996, Calicut, India. pp. 45-48.

22. Shukla, M.R., N. Subhash, D.R. Patel, S.A. Patel and S.J. Eapen, 1997. *In vitro* selection for resistance to Alternaria blight in cumin (*Cuminum cyminum* L.). In: Biotechnology of Spices, Medicinal and Aromatic Plants. (eds. S. Edison, K.V. Ramana, B. Sasikumar and L.N. Babu). Proceedings of the National Seminar on Biotechnology of Spices, Medicinal and Aromatic Plants, 24-25 April 1996, Calicut, India, pp. 45-48.

23. Skirvin, R.M., K.D. Mcfeeters and M. Norten, 1994. Source and frequency of somaclonal variation. Hort Science 29: 1232-1237.

24. Tawfik, A. and G. Noga, 2001. Adventitious shoot proliferation from hypocotyls and internodal stem explants of cumin. Plant Cell, Organ and Tissue Culture 66: 141-147.

25. Tawfik, A.A. and S. Gianinizzi, 1998. Plant regeneration in callus culture of cumin (*Cuminum cyminum* L.). In: Proceedings of the Symposium on Plant Biotechnology as a Tool for the Expoliation of Mountain Lands (eds. S. Scannerini, A. Baker, B.V. Charlwood, C. Damiano and C. Franz), 25-27 May 1997, Turin, Italy, Acta Horticulturea, 457: 389-394.

CHAPTER 8

Economic Aspects

A. Karbasi
Karbasir2002@yahoo.com
Faculty of Agriculture, Zabol University, Iran

8-1 INTRODUCTION

India and Iran are two major producers, and suppliers, of cumin in the world. Considering the unique position of this crop in the world and its specific location across India and Iran, mainly Khorasan in Iran and Rajasthan in India, the economical dimensions of this crop have made phenomenal inroads in the local and foreign markets, the magnitude of which is yet to be fathomed. To obtain a deeper insight of the economy of cumin production, the following features are reviewed in this chapter : (1) economical aspects of cumin at federal level and beyond, (2) distribution process and export. Moreover, the handling and marketing network inside the country, initial and terminal source of export, and problems encountered are being studied. Meanwhile, special attention has been paid to socio-economical characteristics of the producers and the variations in the price of cumin.

8-2 ECONOMICAL VALUE

Cumin cultivation creates significant job opportunities in production areas. Planting, maintenance and harvesting of this crop requires adequate manpower, hence providing considerable employment in the region. On the other hand, since the major portion of cumin produced is earmarked for export, it is a valuable cash crop (7,8).

Table 8-1 shows the quantity and value of cumin produced during 2000–2001 in Khorasan. It shows that the value of crop production in different cities, using the mean price, is a noteworthy total of 97,545 million Rials. A considerable amount of this is attributable to value added results from work power, i.e. the value added from marketing services. The price will be much more, which shows the economic importance of this crop after producing in production areas and the value added due to job opportunities. Therefore, the presence of this crop in cropping pattern can be economically important.

The price of cumin depends on market demand and supply. However, since most of the cumin produced is exported, the export price is determined by the quantum of external demand. The exporters of this crop evaluate the price and market demand, and the farmer receives the price based on the market value.

During recent years there were fluctuations in the price of cumin because of variation in external demands and polarity of internal market. The trend of price variation indicates that with increase in the inflation rate, the price of the crop is high. Since the exported cumin price before 1958 was about 0.2 US $, and during 1959-1960 it was about 0.45-0.5 US $ (1), in the following years the price of cumin has been increasing annually, and in the recent years it has touched 4.0 US $ of cumin. The price of cumin in Rajasthan (India) has been 1.5-2 US $ per kg in last four years. A drop in export demand of cumin results in a considerable fall in price.

8-3 MARKETING

Since cumin is of little household value, and mostly caters to market for sale, it is the marketing strategy that plays the primary role in generating additional profit for the producer. At a glance, the marketing of cumin may be divided into internal market and external market. In the internal market, the network that cumin follows from farm producer to the final consumer is being reviewed. In the external market, usually the final market and pathway that cumin follows is the focal point. A couple of researches have been carried out in the internal department, but no study has been conducted in the export section. As a consequence, efforts have been made to review the internal market and then the export segment may be explored.

Table 8-1 *Quantity of production, value of cumin/kg, and overall price of cumin crop in Khorasan areas of Iran during 2000-2001 growing season (4)*

No.	City	Total cumin production (tons)	Price per kg (Rials) *	Total price (Million Rials)
1	Bardascan	182	15,000	2,730
2	Birjand	182	15,000	270
3	Bojnord	17	15,000	1,755
4	Daragaz	2	15,000	30
5	Esferayen	930	15,000	13,950
6	Fariman	3	15,000	45
7	Ferdows	1,766	15,000	26,490
8	Ghaen	33	15,000	495
9	Gonabad	298	15,000	4,470
10	Jajarm	175	15,000	2,625
11	Kashmar	14	15,000	210
12	Khaf	45	15,000	675
13	Mashhad	61	15,000	915
14	Nehbandan	1	15,000	15
15	Nishaboor	103	15,000	1,545
16	Quchan	11	15,000	165
17	Sabzevar	1,488	15,000	22,320
18	Sarakhs	-	15,000	-
19	Shirvan	10	15,000	150
20	Tabas	22	15,000	330
21	Taibad	336	15,000	5,040
22	Torbat-e-Heyderiyeh	622	15,000	9,330
23	Torbat-e-Jam	266	15,000	3,990
	Total	6,567		97,545

* Current rate is 9100 Rials per US$

8-4   NATIONAL MARKETS

The available markets include: on farm market, wholesale market, and retail market. On farm market is restricted to farms, and the sellers are farmers. This market is controlled partially by some supplier agents. The marketing agents in this market are mainly farmers, stock suppliers, local buyers and representatives of wholesalers, mediators, and market owners. Considering that the representatives of stock suppliers offer a higher price compared to other buyers, they have the higher contribution in cumin commerce on farm; on the other hand, the market owners have the least contribution in cumin exchange (3).

The main exchange of cumin is usually done simultaneously at harvest either inside the farm or in a special place designated in the village. In this market, the exchange is done by agreement and inspection of the product. The seeds are not graded and contain impurities. The on farm exchange is mostly in cash because the buyers and sellers usually do not know each other and they rarely accept short-term or long-term payment.

If a farmer is not willing to display his crop on farm market, he pays for packing and handling, and delivers his product for wholesale in the stock market. The center of the stock market at the city level is usually located at cumin sifting factories. The sellers at stock markets are mostly farmers, and in some cases local sellers and representatives of wholesalers are also present at the market site where the products are displayed. The buyers of cumin stock markets, based on order of priority, are: the stock sellers of other cities, retailers , mediators, and consumers. Parts of the stock exchange market are in the hands of sifting factories. In this market, cumin is usually exchanged via a receipt and a bank draft obtained from factories. The sifted good is exhibited within the factory and the exchange here is similar to that of the stock market, and based on supply and demand. The buyers are mostly wholesalers, and exporters or their representatives.

Almost half of the goods exchanged at the stock market are sold on short- or long-term payment, and the rest of it is in cash payment. In the case of term payment, usually the farmer asks the buyer to pay for about half of his goods in cash, and the balance after the buyer sells the commodity. Stock sellers and wholesalers usually conduct transactions independently, and rarely percentage wise. In the latter case, they generally ask for a percentage payment for selling the crop. However, they are reluctant to agree to this.

If the farmer is not willing to offer his crop to sellers on farm market or stock market, and he is able to spend some funds on advertising and marketing, he can take his product to the city and sell it to retailers at the optimum price, but the value at which the retailer buys it, is not high compared to other markets.

Figure 8-1 shows the marketing pathway of cumin. In this distribution network, the production and marketing corporations, and organizations established by them, are behind the scene, and this has resulted in the farmers being forced to follow the market variations.

8-5 CUMIN MARKETING PROCESSES

The process through which the product passes right from the farm and finally to the end-user, i.e. consumer, involves several steps such as – purification, grading, packing, handling, storage, sifting and conditioning.

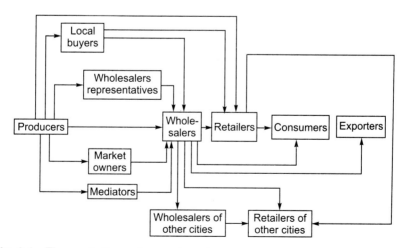

Fig. 8-1   The marketing pathway of cumin.

Coarse purification of the crop was done primarily after harvesting with elementary tools. In such purification, plant debris, soil aggregate and a percentage of other impurities were reduced chiefly by naturally sifting. The higher the purity of the product, the higher is the marketing price. This function depends on the skill, experience and the patience of the farmer. At this stage, some farmers grade the product based on size and color, and select their premium product for planting in the next year, which is usually based on seed size. At this stage the product rarely delivered to the market.

In the next stage, the product is packed in 50 kg bags. Packing may be done on farm or in storage. The packed product will be delivered for further cleaning, or to stock sellers or retailers.

Handling the product from farm to village or to stock seller is usually done by truck or pick up service, and since the seeds are dry, they do not need a special type of vehicle or specific treatment for translocation. Proper storage is essential for maintaining high quality and protecting the product. The stores used for the product in the villages are usually located in farmer's house and at stock sellers premises; they are usual stores which are not up to the requisite standard. Debris and pests are rarely seen in such storerooms, but in some cases spoilage due to improper storage may be observed in the product. Moreover, the period of storing and the type of storeroom is very effective in reducing the weight of the product due to evaporation of partial water from seeds.

The cleaning procedure is carried out in appropriate factories for fine purification. Therefore, the crops planned for export are transferred to cleaning factories and after fine cleaning, the product is ready for further processing and export. In Iran, the cleaning factories are limited in Khorasan and mostly located around Mashhad and Sabzevar.

8-6 EXPORT

Cumin is an important and valuable crop, which due to its special ecological characteristics for cultivation, is planted only in limited parts of the world. India, Iran, Indonesia and Lebanon are the main exporters of cumin, but to a lesser degree countries such as Turkey, Egypt, Syria, China, Cyprus, Argentina, and Mexico also export cumin. The export of cumin from Iran dates back to the 1950s. For instance, in 1958 Iran exported 8,799 tons of cumin. Iran was the number one exporter of cumin until 1960, but ranked number seven in 1970 (1, 9). Most of Iran's cumin export is to Dubai, Pakistan, Japan, Cambodia, Kuwait, UAE, Germany, France, Netherlands, Czechoslovakia, Saudi Arabia , Qatar, Hungary, Singapore, Belgium, and England .

Table 8-2 shows the quantity of cumin exported from Iran to different countries during 1996–2000. Main importers of cumin from Iran :

(Figures in parentheses represent percentage of total cumin exported by Iran)

- 1996: Ukraine   (55%)
          UAE   (33%)
          Russia   (7%)
          Pakistan   (2%)
- 1997: The number of countries importing cumin from Iran reduced from 19 in 1996 to 15 in 1997.
          Ukraine   (77%)
- 1998: Imported by 24 countries
          Ukraine   (46%)
          UAE   (32%)
          Qatar   (7%)
          Japan   (6%)
          Morocco   (1%)
- 1999: Imported by 16 countries
          UAE   (73%)
- 2000: Pakistan   (38%)
          UAE   (33%)
          Ukraine   (13%)
          Japan   (4%)
          England   (3%)
          Germany   (2%)

Table 8-2   *A checklist of 10 most important importers of cumin from Iran and the quantity of cumin exported during 1996–2000 (2)*

(Qty. in Kg)

No.	Country	1996	1997	1998	1999	2000
1	Germany	-	-	2,491	35,715	282,960
2	Uzbekistan	19,838	141,817	100,205	117,619	97,938
3	Spain	22,680	71,064	291,360	119,340	125,334
4	UAE	6,717,740	1,464,660	10,745,572	7,597,249	251,232
5	England	54,000	79,380	285,336	-	265,214
6	Ukraine	11,225,588	4,990,041	7,483,793	1,875,906	998,239
7	Turkey	-	11,662	19,656	186,556	6,6694
8	Pakistan	244,666	-	-	-	2,940,447
9	Qatar	16,250	8,100	1,434,901	158,891	132,373
10	Japan	-	79,380	522,725	-	297,000

Table 8-3 shows the quantity of cumin exported from Iran since 1966 to 2001. It is observed that the export of cumin during 1971–1978 was relatively stable and the rate was increasing to some extent. After the Islamic revolution in Iran, by increasing fluctuation in the market, there was a pronounced decline in the quantity of cumin exported to other countries. From 1989, like other products the export of cumin also increased, and in 1992 it reached a peak of almost 32,000 tons per year. Some restrictions during recent years have resulted in cumin exports declining to a point that during 1995 by fixing the exchange rate and establishing exchange rate convention, the exported quantity reduced to less than 10,000 tons. However, during the period 1971 to 1999, cumin export had an annual growth rate of 4%. Investigations indicate (5) that Iran's contribution from European Economical Union Market is 32.3% and by removing the restriction of export and exchange convention, it may be promoted up to 45%. The reasons for declining cumin exports are as follows:

1. Incompetence between exporting exchange rate in comparison with inflation rate, and the final price of internal production.

2. Lack of control and careful monitoring of the quality of cumin by Standard and Industrial Research Orgnization, Iran.

3. Lack of proper storage conditions and packaging of cumin

(which should conform to international standards) and not considering sanitation and exceptional taste of consumers.

4. Lack of extensive and effective advertising to encourage and support producers and exporters.
5. Lack of requisite infrastructure in the country and inadequate technology for the production and processing of cumin.
6. Emergence of new competitors in cumin production in the world market, using better selection criteria, and advanced and sophisticated tools in cumin production, processing, handling, and conditioning.
7. Exporters not having adequate expertise in export policies and procedures.
8. Illegal border trading, illegal exporting and pushering cumin.
9. Limitation in cumin exports such as obtaining authority and exchange convention to export cumin.

8-7 SOCIO-ECONOMICAL CHARACTERISTICS OF CUMIN PRODUCERS

The number of people who benefit from cumin varies annually at state and federal levels. It is estimated that around 10,000 to 80,000 people are involved in cumin growing during draught or rainy season respectively. These people have typical socio-economical status. Most of the cumin crop grows in Khorasan and researches indicate that the average age of cumin planters is 49 years.

Age of cumin planters :

Age (years)		Ratio
30-40	...	18.9%
41-50	...	24.5%
51-60	...	35%
Above 60 years	...	21.6%

From the above it is apparent that 81.1% of cumin planters are over 40 years of age. The majority of Iran's population comprises the younger generation, but they have not taken up cumin cultivation as a profession.

The level of education in the understudy society was disastrous. 62.6% of the examiners were illiterate, 31.1% could only read and write, 4.8% had a degree in elementary school, junior high school, or senior high school, and only 1.1% had formal university education.

Table 8-3 *Quantity and value of cumin exported from Iran during 1966-2001 (2)*

No.	Year	Quantity exported (Tons)	Value (Million Rials) *	Value (Million US$)	Average value (US$/Ton)
1	1966	9,900	374	-	-
2	1967	9,355	266	-	-
3	1968	12,080	244	-	-
4	1969	13,260	248	-	-
5	1970	8,565	274	-	-
6	1971	4,535	193	-	-
7	1972	7,567	305	-	-
8	1973	4,365	268	-	-
9	1974	5,886	416	-	-
10	1975	4,234	305	-	-
11	1976	5,969	404	-	-
12	1977	6,465	456	-	-
13	1978	6,290	522	-	-
14	1979	2,961	299	-	-
15	1980	1,037	116	-	-
16	1981	225	30	-	-
17	1982	2,219	361	-	-
18	1983	1,779	313	-	-
19	1984	903	116	-	-
20	1985	894	99	-	-
21	1986	7,638	669	-	-
22	1987	5,664	549	-	-
23	1988	3,482	595	12.22	112.0
24	1989	10,864	856	18.53	1,073.0
25	1990	14,058	1,298	27.53	1,139.0
26	1991	24,162	1,930	36.12	1,140.0
27	1992	31,428	2,533	36.12	1,140.0
28	1993	23,438	1,888	26.87	945.0
29	1994	19,435	38,177	24.45	1,155.5
30	1995	9,666	18,100	10.34	1,069.7
31	1996	20,270	35,654	18.39	922.9
32	1997	6,810	11,957	6.08	1,004.9
33	1998	21,378	38,403	21.90	1,036.6
34	1999	10,429	19,989	11.30	1,082.4
35	2000	7,881	15,095	8.60	1,091.3
36	2001	5,089	11,216	6.40	1,255.8

* Current rate is 9100 Rials per US$

Most cumin planters in the understudy society have small farms. In this society 47.1% of farmers had 0.5 hectares acreage. 24.4% owned 0.5 to l hectare land and 25% between 2 and 5 hectares, and only 3.5% had 5 to 10 hectares cumin plantation. Therefore, it may be concluded that most cumin planters in Khorasan are aged farmers with no education or with low levels of education. Hence, not possessing the required skills, results in increase in production expenses and inability to overcome the limitations and problems encountered by farmers, leading to undesirable consequences in cumin development (6).

8-8 SUMMARY

Iran is one of the major producers of cumin in the world and approximately 60% of world cumin export belongs to Iran. The cumin acreage and production has increased during the recent years. Cumin plantation extended to several states across Iran. Khorasan is the main cumin producing area in the country and its production plays an important role by creating job opportunities and value addition. The marketing practices of the crop are carried out in traditional markets. Considering that most of the cumin is exported to different countries, not too much processing is done locally on this product. The quantity of exported cumin varied during several years and resulted in high fluctuation in the market price. The presence of some limitations and problems resulted in inconsistent development of cumin exports. Cumin is exported to several countries, but during recent years, UAE, Pakistan, Japan, Germany, and Netherlands are the main importers of Iran's cumin, in particular, UAE and Pakistan imported the bulk of Iran's cumin volume recently. Since the climate of Iran is favorable for cumin growth and due to Iran's long history of cumin exports, a hopeful perspective for more extensive contribution of Iran in the world cumin market is predicted. However, in spite of having premium quality, this product is suffering due to lack of appropriate marketing strategy. A suitable policy and procedure should be drawn up for better marketing of the crop locally and abroad.

REFERENCES

1. Balandari, A. 1991. The study of botanical characteristics of endemic cumins. Iranian Science and Technology Research Organization, Khorasan Research Institute, Mashhad, Iran.

2. Custom external business annuals. Different years. Tehran, Iran.

3. Karbasi, A. 2002. The Marketing study of cumin and barberry. Research Project of Ministry of Jahad Agriculture, Institute of Agricultural Economics and Programming Research, Tehran, Iran.

4. Khorasan Jahad Organization, Programming and Design Management, Internal annual Reports, Mashhad, Iran.

5. Kimjani. A. 2001. Rules and regulation of accessing to agricultural crop market and its effect on Iran agricultural economics, Agricultural Jahad Publications, Tehran, Iran.

6. Mohammadi, F. 1996. The study of saffron and cumin Production and exportation. Agricultural Economics and Development. Selected Articles of Seminar. Tehran, Iran.

7. Molla Filabi, A. 1992. The effect of planting dates and row spacing in cumin yield under irrigated and dry land conditions. Iranian Science and Technology Research Organization, Khorasan Research Institute. Mashhad, Iran.

8. Rahimian, H. 1990. The effect of planting dates and irrigation on cumin yield. Iranian Science and Technology Research Organization, Khorasan Research Institute. Mashhad, Iran.

9. Sadeghi, B. and M. H. Rashed-Mohassel. 1991. The effect of Nitrogen and irrigation in cumin production. Iranian Science and Technology Research Organization, Khorasan Research Institute, Mashhad, Iran.

CHAPTER 9

Processing, Chemical Composition and Standards of Cumin

A. Hemmati Kakhki[1] and M. Sanuie Mohassel[2]
[1] hemati@kstp.ir
Iranian Research Organization for Science & Technology, Khorasan Center, Mashhad, Iran
[2]-Exir@kstp.ir
Nader Production and Processing Campany, Mashhad, Iran

9-1 INTRODUCTION

Cumin, since the ancient days, has been used by the people of Egypt and India as the main spice for different kinds of dishes, the chief reason being its unique aroma and exceptional taste. Nowadays people prefer to use natural products rather than chemicals and artificial flavors. Cumin is used mainly where highly spiced food is preferred. It is an ingredient of most curry powders, and many savory spice mixtures (1, 15, 16, 18). Cumin is also a medicinal plant and in many pharmacopoeias it is introduced as a medicinal herb because of its antibacterial, stomachic, diuretic, carminative, stimulant and antispasmodic properties. In the same manner, cumin can be used in the cosmetic industry (14, 16, 18).

Although cumin is cultivated in many countries around the world, four countries, i.e. India, Iran, Indonesia and Lebanon are the main exporters of this spice seed (9, 15).

9-2   CHEMICAL COMPOSITION OF CUMIN

9-2-1   General

Fresh seed of cumin is greenish yellow in color, but it may turn brown or gray based on the time of harvesting and duration of storage. Seed pericarp is full of tannins and in the presence of ferric salts, its color changes (17, 18).

Cumin seed comprises around 7% fat, 13% resins and 2.5 - 4.0% essential oils. The total ash content of cumin seed should be less than 9.5%, non-soluble ash in acid less than 2.0% and moisture content less than 9%. Total non-toxic external materials in cumin mass must be less than 0.5%. Crude protein, true protein, non-protein nitrogen and amino acid composition were estimated in cumin seeds supplied by Bulgaria, Egypt and Turkey for two seasons. The Bulgarian cumin had most crude protein (23%) and the Egyptian seeds the least (18%). Generally, 18 amino acids were identified in all cumin seeds, of which eight were essential amino acids. The biological value of the protein was calculated; the first limiting amino acid in cumin was tryptophan (3). Table 9-1 shows the maximum chemical content of 100 grams of cumin seed (9).

Table 9-1   *Maximum chemical content of 100 gram of cumin seed (9)*

Chemical	Amount	Unit	Chemical	Amount	Unit
Water	8.1	g	Phosphorus	449	mg
Energy	375	KCal	Potassium	1788	mg
Protein	17.8	g	Sodium	168	mg
Fat	22.3	g	Zinc	5	mg
Total carbohydrate	44.2	g	Ascorbic acid	8	mg
Ash	7.6	g	Thiamine	1	mg
Fiber	10.5	g	Niacin	5	mg
Calcium	931	mg	Vitamin A	1270	International
Iron	66	mg	Other vitamins	Negligible	-
Magnesium	366	mg			

9-2-2 Oleoresin of Cumin

Extracted oleoresin from cumin seeds by organic soultants, include resin compounds other than essential oil, which is the raw material in food and other industries. This compound is brown or greenish yellow in color and each 100 g of oleoresin contains around 60 ml of pure essential oil. Effective content of each 5 kg of cumin oleoresin has the same value of 100 kg of cumin seed (9).

9-2-3 Cumin Essential Oil

There are seven ridges and oil canals in cumin seeds consisting of essential oil. Physical characteristics of cumin essential oil are as below:

Extractable essential oil	2.3-5.7%
Color (fresh oil)	colorless or bright yellow
Refractometric index (at 20°C)	1.47-1.50
Photo-rotation (at 25°C)	+4° 6′ - +3° 8′
Density (at 20°C)	0.90- 0.94
Solvability in alcohol 80% (vol./vol.)	0.33- 0.50

Guenther (10) pointed out that the physico-chemical properties of cumin essential oil are as per details given below:

Refractometric index (at 20°C)	1.494-1.507
Photo-rotation (at 25°C)	+3° 20 - +8°
Density (at 20°C)	0.90- 0.93
Solvability in alcohol 80% (vol./vol.)	0.1- 0.3
Cuminaldehyde (phenylhydrazin method)	0.35-0.42

In another experiment, cumin seeds from the Mediterranean region showed the following characteristics:

Refractometric index (at 20°C)	1.5011-1.5038
Photo-rotation (at 25°C)	+4° 22′ - +5° 6′
Density (at 15°C)	0.917- 0.924
Solvability in alcohol 80% (vol./vol.)	0.2- and more
Cuminaldehyde (hydroxylamine hydrochloride method)	47.4-51.5 %

Many factors such as plant variety, area of cultivation, agronomic conditions, date of harvesting, method of essential oil extraction and storage conditions of seed and essential oil, could affect physico-chemical properties of cumin essential oil. Volatile seed oil contents determined by distillation of ground 100 g samples of Indian cumin ranged from 2.3% (UC209) to 4.8% (UC198). The correlation coefficient between yield and volatile oil content had slightly higher negative values in cumin (1). Table 9-2 shows the essential oil content of cumin from different parts of the world (15).

Table 9-2 *Essential oil content of cumin seeds from different parts of the world (10)*

Country	Essential oil (%)
India (East)	2.3- 3.5
Iran	2.1-3.0
Malta	3.5-5.0
Morocco	3.0
Syria	2.5- 4.0

9-2-4 Chemical Characteristics of Cumin Essential Oil

Cumin chemical and physical characteristics are affected by the environmental and genetic conditions. Some average chemical characteristics of cumin essential oil are shown in Table 9-3.

Table 9-3 *Average chemical characteristics of cumin essential oil*

Compound	Amount	Unit
Cumin aldehyde	35-63	%
pH (cuminic acid)	0.36- 1.8	Without unit
Cumin alcohol	.3.5	%
Carbonyl index	9.32	%
Ester index	19.24	%

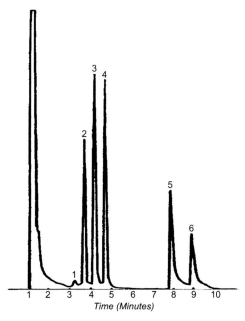

Time (Minutes)

Fig. 9-1 Chromatogram: Analysis of cumin essential oil obtained from gas chromatography.

Karim *et al.* (13) analyzed the essential oil of cumin seed using gas chromatography and reported that there are six common compounds which exist in all the varieties they examined. The chromatogram of these compounds is shown in Fig.9-1 and the type and amount of cumin essential oil components is shown in Table 9-4 (13).

Table 9-4 *Type and amount of cumin essential oil components by gas chromatography (13)*

No. of pick in Fig. 9-1	Compound	% Of total	Range
1	α- penine	1.0	0.5-1.3
2	β- penine	18.7	8.8- 20.0
3	Cumin aldehyde	27.3	26.5-62.0
4	Para-cymene, limonene, phellandrene,cineole	26.0	-
5	Gamma- terpenine	12.2	9.5-30
6	1, 3-papa menthandial -7-al, 1, 4 para mentadiene -7-al	14.4	-

Baser and co-workers (5) analyzed steam and water distilled Turkish cumin seed oils by a combination of GC and GC/MS. The major compounds in the oils are shown in Table 9-5. The mode of distillation and pre-grinding of the seeds had a pronounced effect on the composition of the resultant oil.

Table 9-5 *Type and amount of Turkish, Indian and Egyptian cumin seed oil components by gas chromatography and mass spectrometry (17)*

Compound	Range (% of total)
- penine	2.98-8.90
Cumin aldehyde	19.25-27.02
Para-cymene	4.61-12.01
Gamma-terpenine	7.06-14.10
1, 3-para menthandial -7-al,	4.29-12.26
1, 4 para mentadiene -7-al	24.48-44.91

The influence of particle and batch size, and distillation rate on cumin seed oil extraction was investigated. The major components of cumin essential oils were cumin aldehyde (27.60%), gamma-terpenine (17.25%), p-mentha-1,3-dien-7-al (15.18%), betapinene (10.22%) and p-mentha-1,4-dien-7-al (9.48%) (6).

In addition to the above mentioned compounds, there are many more compounds present only in insignificant quantities, amongst which, cuminile alcohol, cimonene, phellandrene, cineole, 1, 3, -P-mentadiene, champhene, myrecene, myrtenal, capa cymene, dihydro-cuminaldehyde, caryophillin, á–terpineol, linalool, terpinolene, Pulegone could be mentioned (4, 12).

Cumin aldehyde

The most common and most important compound, which comprises approximately 63% essential oil of cumin, is aldehyde cuminal or cumin aldehyde.

The existence of this compound in cumin, with the general formula of $C_{10}H_{12}O$ and chemical name of Para-isopropile benzaldehyde, was reported by Bertanini for the first time and after that Kraut extracted and purified bisulfite salt of this compound (11).

The molecular weight of cumin aldehyde is 148.30 g and its chemical structure shown in Fig. 9-2.

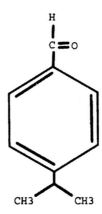

Fig. 9-2   Chemical structure of cumin aldehyde

The presence of cumin aldehyde in cumin essential oil can be determined by methods using : (1) bi-sulfite, (2) phenyl hydrazine, and (3) hydroxide amin hydrochloride. Boiling point of cumin aldehyde in 7mm mercury pressure is 97-99°C, with a density of 0.9731, deviation angle of +3, melting point of 55-57°C. This compound changes to cuminic acid due to oxidation in the presence of crusic acid (6). The cumin aldehyde content of fresh seeds is markedly more than old seeds. Due to the phenolyc structure of cumin aldehyde it has antioxidant activities, and the antioxidant property of cumin essential oil is due to this compound (2). The unique aroma of cumin seed and its essential oil is also because of cumin aldehyde.

Betapinene

The general formula of betapinene is $C_{10}H_{16}$ with molecular weight of 136.23 g. It is produced due to oxidation of a left ventricular acid i.e. nopinic acid, with a melting point of 125°C. This acid is oxidized by potassium permanganate and the result is nopinone, which is a toxic salt, carbazone, with melting point of 187°C.

Para-cymene

Para-cymene is one of the main components of cumin essential oil. Due to oxidation of para-cymene, hydroxy iizopropile benzoic acid with a melting point of 155-156°C is produced. This compound shows the maximum absorption in methanol at 258 and 162 nm.

Cuminile alcohol

Cuminile alcohol, $C_{10}H_{14}O$, with chemical name of P-isopropyl benzyl alcohol and molecular weight of 150.21 g, is produced from a high melting point at 100 to 115°C. In spite of cumin aldehyde, this acid does not react with bisulfate, but its smell is similar to that of cumin aldehyde. When it is oxidized by potassium permanganate, the result is cuminic acid with a melting point equal to 112°C (13).

9-3   PROCESSING OF CUMIN

Before essential oil extraction, cumin seeds should be prepared for distillation. Firstly, the seeds are separated from the whole plants, and all external materials, particularly plant tissues, removed from the seed mass. Essential oil which is extracted from non-cleaned cumin seeds does not have the superior quality. In order to facilitate steam penetration and essential oil extraction, seeds are crushed after cleaning. Crushed seeds are transferred to the silage, but the time between crashing and essential oil extraction should be minimized (6, 14).

9-3-1   Cumin Essential Oil Extraction

This process is done in the evaporation chamber, a cylindrical tank, which has a diameter nearly equal to its elevation, and contains a reticular plate inside. The evaporation chamber has a writhen upper end connected to a tube that transfers essential oil and distilled water in the form of gas to the condenser (Fig. 9-3).

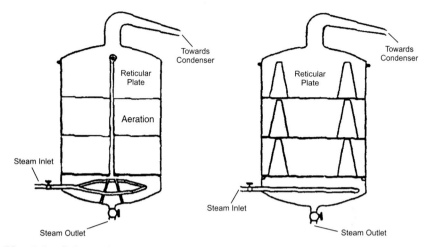

Fig. 9-3 Schematic shape of evaporation chamber with reticular plates and meshed coil.

In the evaporation tank, crushed seeds are placed in the path of steam and essential oil is extracted, which moves with steam to the condenser. Liquidation of steam is done in the condenser.

There are two types of condensers:

1. Coil condenser
2. Vertical tube condenser

In both the condensers a series of narrow tubes are located in a cold water container and when the steam passes out from the steam chamber trough these tubes, it gradually condenses into liquid (Fig. 9-4). The direction of steam movement in the tubes is opposite to that of the cold-water movement.

There are many methods for extraction of essential oil from plant materials. Selection of the best method depends upon the type and amount of essential oil in the plant material, location of reservation in the tissue, economic value of the essential oil, type of plant tissue (fresh or dry) and many other factors. Generally, there are five main methods for extraction of essential oil, which are:

1. Distillation method
2. Solvent method

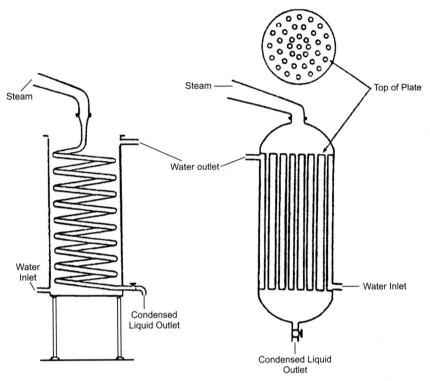

Fig. 9-4   Schematic shape of coil and vertical condenser.

3. CO$_2$ method (7)
4. Physical and mechanical methods
5. Microwave method (6, 10, 15).

The most common method for cumin essential oil extraction is distillation (4, 6). In this method the heat of the steam forces the tiny intercellular pockets that hold the essential oil to open and release the oil, and after condensation and cooling, the essential oil can be separated from other liquids.

Distillation method itself is divided to three sub-methods consisting of: (1) distillation by water, (2) distillation by water and steam, and (3) distillation by direct steam. The basics of all three methods are similar, except application of water steam to the plant material.

However, the extracts obtained using supercritical carbon dioxide at 40°C and 100 bars were compared with cumin essential oil obtained by conventional steam distillation, using GC-MS. The analyses show that valuable components are extracted by the supercritical procedure, which would otherwise be thermally degraded by conventional steam distillation (8).

9-3-1 Water Distillation

In this system, plant tissue is put into the boiling water and the resulting steam condenses. By application of this method, the insoluble cumin essences are separated and collected in the form of a film above the water. This method is simple and does not need any high investment but the quality of extracted essential oil is inferior to that obtained by other methods (Fig. 9-5).

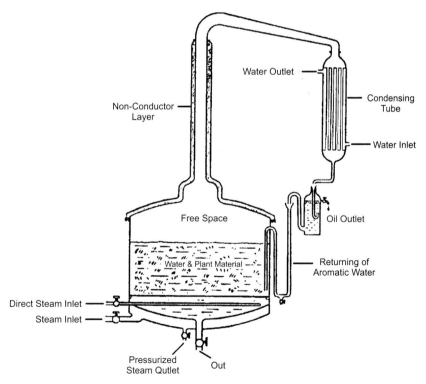

Fig. 9-5 Extraction of Essential Oil by Water Distillation System

9-3-1-2 Schematic shape of Water and Steam Distillation

In this method, the plant material is placed on the reticular plate well above the level of boiling water in the distillation container. The resulting steam passes through the plant material and causes the release of essential oil from the plant and moves with the steam to the condenser. In this method, plant materials do not come in direct contact with boiling water (Fig. 9-6).

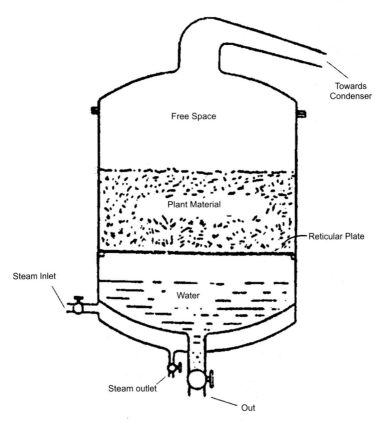

Fig. 9-6   Extraction of essential oil by water and steam distillation

9-3-1-3 Schematic shape of Direct Steam Distillation

In this method, the plant material is placed on the reticular plate in the distillation container while there is no water at the bottom of the still. Fresh and sometimes dried pressurized steam is generated in a separate container, and is usually pumped at a pressure which is higher than atmospheric pressure, which enters through a meshed coil to the cumin seed or crushed seeds container. The force of the steam passes through the plant material and causes it to release essential oil from the plant material and moves with the steam to the condensation chamber through a tube from the top of the still. (Fig. 9-7).

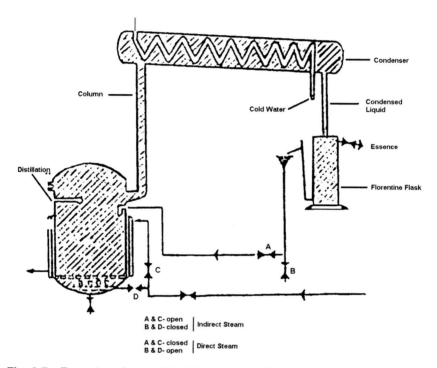

Fig. 9-7   Extraction of essential oil by steam distillation.

In all distillation methods, essential oil after condensation is mixed with some water, but in two separate layers, cumin essential oil floats on the top upper layer and the water underneath, which is in fact called floral or aromatic water, and contains some essence. In order to increase the efficiency of essential oil extraction system, this water is sent back to the distillation system to re-extract its remaining essences. In this system, separation of essential oil is done using a florentine flask.

9-4 SCHEMATIC SHAPE OF SEPARATION OF ESSENTIAL OIL

In this stage, essence and condensed water are separated based on their weight difference, resulting in two separate layers. In the florentine flask, essence and water are not separated easily. Therefore, condensed liquid should enter into the separator slowly to prevent any turbulence in the liquid.

9-4-1 Standardization of the Essence

After the separation of essential oil, the flask should be kept still to let the small droplets and other external materials separate. The essential oil then passes through special filters and columns. In these columns different components of the essential oil separate based on their boiling points.

9-5 PACKAGING

Cumin essential oil should be kept in dark and airtight containers. The suitable containers for storage of essential oil are aluminum containers, colored glass bottles and epoxy-coated barrels. Cumin essential oil should be stored in a cool and dry place.

9-6 CUMIN STANDARDS AND QUALITY CONTROL

Cumin being a popular spice should fulfill at least the prescribed standards of the country of origin to be acceptable in the international market. At international level, International Standard Organization is responsible for ensuring standards for each product and at national level each country establishes its own standards. International Organization for Standardization is a worldwide federation of national standard bodies. A technical committee, Agricultural Food Products Subcommittee for spices and condiments, prescribes ISO 6465 for cumin. The ISO 6465 consists of specifications and methods for conducting tests. Chemical characteristics of cumin based on international standard, ISO 6465 are represented in Table 9-6.

Table 9-6 *Classification of cumin based on Iranian standards number 14 and European Spice Association (ESA)*

Characteristics	Categories			
	I	*II*	*III*	*ESA*
Moisture % (m/m) max.	9.0	9.0	9.0	13
Total ash % (m/m) on dry basis max.	9.50	9.75	10.0	14
Acid insoluble ash % (m/m) on dry basis max.	1.50	1.75	2.00	3
Non-volatile ether extracts % (m/m) on dry basis min.	15.0	15.0	15.0	-
Essential oil ml/100 g dry basis min	2.50	2.45	2.4	1.5

9-6-1 Cumin Standards in Iran

More than 80% of Iranian cumin is exported mainly to Pakistan, United Arab Emirates, Ukraine, Japan, Germany and England. It is, therefore, important that the quality of the product matches national and international standards.

Institute of Standards and Industrial Research of Iran (SIROI) is responsible for ensuring the quality standards of the products, including cumin. The main objective of cumin standard is to specify the norms for packing, labeling, sampling and methods of testing, and whether the particular produce is fit for seed, powder form, or essential oil. In this standard after definition of terms, specification of standard cumin was introduced and published under the title 'Cumin Specifications' under registration number 14 (11). It is applicable to whole seed and in powder form. In this standard, cumin is classified into three groups (Table 9-6) based on essential oil content and extraneous matters.

There is another section of standard, which introduces the methods suitable for testing cumin and it is applicable to seed and cumin powder. In this part, besides explaining general tests such as moisture and extraneous matter, blackened and insect bored seeds, determination of total ash and acid-insoluble ash is also included. This publication in the name of "whole cumin specification" under number 14 was released by the Standard and Industrial Research Organization of Iran (11).

9-6-2 Cumin Standards in India

Cumin standards under the title of "spices and condiments - cumin (safed jeera) whole- specification" adopted by the Bureau of Indian Standards after the draft finalized by the Food and Agricultural Division Council under the number Indian Standard (IS) 2447, 1993 (reaffirmed 1998). Based on this Act, cumin should be free of any visible insects and moulds and has the below mentioned specifications (7):

1. Total ash content is not more than 8% of the total weight.
2. Acid insoluble ash should not be more than 1.0% by weight.
3. Cumin seeds shall be dried fruit of the plant *Cuminum cyminum* L.) and its moisture content should be less than 10%.
4. Foreign materials such as soil, sand, leaves, stem and straw should be less than 2.5% by weight.
5. The damaged, discolored, weevilled, shriveed and immature cumin seed should not be more than 8.5%.
6. It shall be free from visible mould or insect infestation and musty odor. It shall also be free from any harmful foreign matter.
7. It shall have the characteristic shape, color, taste and aroma normal to the species.

There are three grades of cumin, the characteristics of each group are indicated in Table 9-7.

These rules shall apply to cumin seeds produced in India and shall be in addition to and not in derogation of the General Grading and Marking Rules, 1937.

Cumin seed (powdered) shall be the material obtained by grinding dried cumin seed. It shall be free from admixture, mould growth, insect infestation or musty odor.

Table 9-7   *Cumin seed grading based on Indian rules; these measures are maximum permitted amount (7)*

Category	Extraneous matter % by wt **	Other seeds % by wt***	Damaged discolored and weevilled seed % by wt.****	Shriveled and immature seeds % by wt *****
I	1.0	0.5	1.5	1.5
II	3.0	1.0	2.5	3.0
III	5.0	1.5	5.0	4.0

*Grade: As agreed between buyer and seller

**"Extraneous matter" includes dirt, dust, stone pieces, stalks, stems or any other impurity.

"Other Seeds" include seeds other than that of cumin.

***"Damaged and discolored seeds" are those seeds that are internally damaged and discolored. These affect the quality.

****"Weevilled seeds" are those seeds that are partially or wholly bored, or eaten by weevil or other insects.

*****"Shriveled and Immature seeds" are those seeds that are not properly developed.

Table 9-8   *Cumin powdered grading based on Indian rules; these measures are maximum permitted amount (7)*

Grade Designation	Moisture percentage by weight maximum	Total ash percentage by weight (maximum)	Acid insoluble ash, percentage by weight (maximum)
Standard	12.0	8.0	1.5
General	12.0	9.5	1.5

9-6-3 Adulteration in Cumin Essential Oil

The main type of adulteration in cumin essential oil is addition of synthetic cumin aldehyde to original essential oil. This type of adulteration is not detectable by chemical tests. Since addition of synthetic chemicals changes the light dissociation angle, this is the best method for detecting essential oil adulteration (15, 18).

9-7 SUMMARY

Cumin is a valuable medicinal plant and has many applications in the food and pharmaceutical industries. In addition to many normal components of cumin such as proteins, carbohydrates, fats, different elements and vitamins, this plant contains 2.5-4.0% essential oil, which has many uses. The most important components of cumin essential oil are cumin aldehyde, gamma-terpenine, beta pinene, para cymene and alfa pinene. The process of essential oil extraction consists of crop harvesting, seed cleaning, seed crashing, extraction and evaporation of essence, separation, standardization and packaging of essential oil. In addition to consumption of cumin in food and pharmaceutical industries, recently cumin has been introduced as a natural antioxidant material and countries that produce this plant can replace the synthetic antioxidants with the natural essential oil of this plant.

REFERENCES

1. Agrawal, S., R.K. Sharma. 1990.Variability in quality aspect of seed spices and future strategy. Indian Cocoa, Arecanut and Spices Journal. 1990, 13: 4, 127-129.

2. Anon, A. 1985. Extraction of antioxidants from spices, Chem. Abs. 102.

3. Badr F. H., E.V. Georgiev 1990. Amino acid composition of cumin seed (*Cuminum cyminum* L.). Food Chemistry. 38: 4, 273-278.

4. Bandoni, A., M. Juarez and I. Mizrahi. 1991. Essential oil of cumin (*Cuminum cyminum* L.) ESS. Der. Agr. 61:32-49.

5. Baser K.H.C.; Kurkcuoglu M.; Ozek T. 1992. Composition of the Turkish cumin seed oil. Journal of Essential Oil Research. 4: 2, 133-138.

6. Beis S.H., N. Azcan, T. Ozek, M. Kara, K.H.C. Baser 2000. Production of essential oil from cumin seeds. Chemistry of Natural Compounds. 36(3): 265-268. Translated from Khimiya Prirodnykh Soedinenii (2000) 36(3): 214-216 (Ru).

7. Bureau of Indian Standards. 2002. IS 2447. Spices and condiments – Cumin (Safed Jeera) whole- Specification. Second Revision.

8. Eikani M.H.; Goodarznia I.; Mirza M. 1999. Supercritical carbon dioxide extraction of cumin seeds (*Cuminum cyminum* L.). Flavour and Fragrance Journal, 14: 1, 29-31.

9. Farrell, K.T. 1985. Spices, condiment and seasoning. AVI Press. New York. Pp. 97-100.

10. Guenther, E. 1975. The essential oils. Vol. 1-VI, rieger, R.E. Pub. Malabar, Florida.

11. Institute of Standards and Industrial Research of Iran.1996. ISIRI 14, Whole Cumin specification, Third revision.

12. Jiao, X. and J. Sun, 1990. Studies on chemical composition of volatile oil from seeds of *Cuminum cyminum* L. Acta Botanica Sinica. 32: 372-375

13. Karim. A., M. Pervez and M.K. Bhatty, 1985. Studies on the essential oils of the spices of the family *Umbelliferae*, Part III. *Cuminum cyminum* L. Pak, J. Dci. Ind. Res. 19: 229-242.

14. Maskuki, A. 1998. Technology for essential oils and production from aromatic plants. Iranian Research Organisation for Science & Technology, Khorasan Center.

15. Omidbaigi, R. 1998. Approaches to production and processing of medicinal plants (Vol. 2). Tarrahan Nashr publishers (in Persian). Tehran, Iran.

16. www.drmarkvinick.com/articles.htm.2004. The therapeutic properties and usage of five common spices.

17. www.indianspices.com/html/agmark_seed.htm.

18. Zargari, A. 1988. Medicinal Plants. University of Tehran Press. Tehran, Iran.

CHAPTER 10

Research Strategies

M. Kafi[1] and A. Mollafilabi[2]
m.kafi@ferdowsi.um.ac.ir
filabi@kstp.ir
[1]Faculty of Agriculture, Ferdowsi University of Mashhad, Iran
[2]Scientific and Technological Research Organization of Iran, Khorasan Center, Iran

10-1 INTRODUCTION

During the last two decades, research on cumin has been mainly conducted by the agricultural section of Khorasan state branch of Technology and Science Research Organization of Iran, and elucidated some unknown dimensions of cultural practices of this crop. However, in due course of time the ecology, morphology, phonology, physiology, and cytology of cumin will be better understood. In India, Rajasthan Agricultural University and research centers affiliated to Indian Council of Agricultural Research, conducted many research projects regarding control of cumin diseases, introduction of new varieties and agronomical practices.

Cumin is probably one of the scarce medicinal crops for which no single comprehensive book has yet been published in the world that can be used as the chief source of information on all aspects of cumin. Since the producers of cumin are underdeveloped countries, and consumers from other countries obtain the grains with optimum price, there is no an intensive program to promote the efficacy of this product.

Considering the increased input and labor costs of cumin production and processing, there is no economical explanation for production of traditional crops such as cumin, and they are being replaced by breeder crops with high efficacy.

This outcome has resulted in weakening the culture of agriculture in these areas and elimination of biodiversity and disturbance of sustainability of the agricultural systems. The new crops after some period of time, due to high input and susceptibility against adverse environmental conditions, pests, diseases, and weeds, lose their economic efficacy. Therefore, the logistic method is to conduct basic and applied researches for increasing efficiency of endemic and low input indigenous crops. Since these crops have relatively higher water use efficiency (WUE), increasing fertility and yield do not necessarily mean that they need more input and water.

10-2   TREND OF PRESENT RESEARCHES

Although cumin is an important component of traditional and industrial medicine in developed countries, and since this valuable grain is easy to obtain from underdeveloped exporter countries, few researches have been carried out on the ecophysiology and cultural practices of cumin. In an analysis of registered articles in Agricola website, which is an internationally known site where every year more than 100,000 scientific articles are registered in it, only 19 articles on cumin were put up from 1968 to 1997, or a mean of two articles every year. Fig. 10-1 shows the frequency distribution of these articles. The interest of researches and scientific journals in cumin also did not follow a specific trend, and only during 1971 to 1975 the number of articles exceeded the average. Based on these statistics, the relative contributions of published articles in proportion to total registered articles are also shown. In the best years, we have one article on cumin out of 20,000 articles in Agricola. In some years this ratio is one article on cumin out of 100,000 articles.

In an investigation, cumin articles in *Biological Abstract Journal* during 1983 to 2000 were analyzed. There were 81 articles on cumin during 17 years, most of which have been published in cumin producing countries. Figure 10-2 indicates the temporal distribution of these articles. More than 70% of the articles have been published during the 1990s, with an average of about five articles per year.

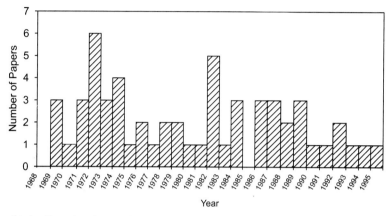

Fig. 10-1 Trends of published papers relating to cumin [extracted from Agricola website (1968-1997)].

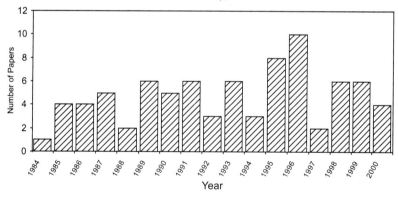

Fig. 10-2 Trend of published articles relating to cumin in Biological Abstracts (1984-2000).

In Iran, relatively extensive researches have been conducted on cumin. Most of these researches were in Iranian Technology and Science Researrch Organization, Khorasan Center. They have conducted 21 projects on this plant (Table 10-1). However, Agricultural Colleges and Agricultural Research Centers in the Ministry of Agriculture had also done some research activities on

Table 10-1 *Projects conducted by Iranian Technology and Research Organization, Khorasan Center, relating to cumin during the last two decades.*

1- Balandari, A.	A botanical survey of indigenous cumin's in Iran
2- Balandari, A. and Rezvani, P.	Comparison of local varieties of cumin in Iran
3- Balandari, A.	The effect of freezing and moisting seeds on germination, growth and yield of cumin
4- Fottowat, A.	The study of nitrogen, phosphorus, and potassium (NPK) on cumin yield under irrigated condition
5- Fottowat, A.	The effect of soil salinity on cumin yield
6- Haddad Khodaparast, M.H.	Designing and construction of cumin distillation
7- Hagian, M.	Chemical control against *Alternaria* blight of cumin
8- Hagian, M.	The study of some fungicide on growth against cumin blight in Khorasan
9- Kafi, M.	The study of different amount of seeds on yield of cumin under dry land and irrigating conditions
10- Kafi, M.	The effect of different amount of planting cumin seeds on yield of cumin under dry land condition.
11- Koocheki, A.	An investigation on cumin production in Khorasan
12-Modarress Razavi, M.	Cumin harvester apparatus
13- Mollafilabi, A.	The study and comparison of different control methods of cumin planting under irrigated and dry land conditions
14- Molafilabi, A.	The study of different method of plantations and amount of seeds under moist and dry planting
15- Mozaffari, G.	The evaluation of yield with physiochemical properties of water and soil plantation of Kashmar
16- Naseri, M.T.	The study of Nitrogen, phosphorus requirements of cumin by Critical concentration method
17- Rahimian, H.	The study of planting dates and amount of seeds on cumin growth and yield
18- Rahimi, M.	The study of chemical control against weeds in cumin crop.
19- Rahimi, M.	The cumin planter apparatus
20- Sadeghi, B.	The effect of different amount of N_2 and irrigation of cumin production
21- Seadatmand, J.	The effect of cumin pulp in cow lactation potential

cumin. The results of these investigations have not yet been applied in the field.

Apparently, a number of problems are being faced in cumin production, the trend of which is much faster than finding the right solutions. During the last few decades, the cumin production per area not only increased, but also witnessed a decrease in production in some regions.

10-3 OBSTACLES

Although green cumin cultivation generates tremendous job opportunities and finance for farmers around desert areas in Iran and India, it is known to them only as a supplementary crop with little significance. The main reason is that regional consumption of this crop is not high and, therefore, reducing cumin production does not cause any effect. However, the fact is that cumin is mainly exported for medicinal purposes and exporting 10 kg of cumin is equivalent to a barrel of petroleum. Therefore, a continuous programmed research is imperative to understand and solve the problems, for the benefit of both the national economy and producers.

Following are the main research channels proposed for increasing cumin production:

10-3-1 Introduction of New Cultivars

After several years of classical plant breeding, Iran is now applying different methods in breeding field crops and horticultural crops, but still does not possess a certified cumin variety produced by scientific centers in the country. Whereas, in countries such as India and Egypt (3, 4) commercial varieties of cumin have been introduced, but in Iran export varieties have not yet been tested. On the other hand, this plant has serious drawbacks that cannot be solved, except for getting little help from plant breeding. Hence, it is essential to collect and identify internal and external cumin genotypes primarily by selection, and then by crossing, introduce appropriate varieties to cumin producers.

10-3-2   Ecophysiological Characterization

Physiological mechanisms of cumin behavior in different environments are unknown. Cumin has a very low leaf area index, with branched and to some extent succulent leaves. Leaf color is light green and plant flowering is photoperiod sensitive. Contribution of different organs to yield is not fully understood. Therefore, basic research on ecophysiological aspects of cumin growth and development is required for both agronomists and plant breeders for development of new high yielding stress tolerant cultivars.

10-3-3   Place in Crop Rotation

Cumin is usually produced in rotation with summer vegetables. However, the probable allelopathic effects of its residues on the following crops are not clear and should be studied in detail.

10-3-4   Cropping Practices

Many researchers have studied agronomy of cumin, but there are conflicts among the results published. For example, there are different views regarding irrigation and fertilization of this species. Some experimental results indicated that cumin has no special irrigation or fertilization requirement and can survive on soil water and nutrient reserves (9). However, other researchers have emphasized the economic importance of irrigation and fertilizers on cumin yield (2). It seems that these conflicts originated mainly from differences in weather and soil conditions. Optimal agronomic practices for ensuring high yields require regional studies based on detailed information about soil characteristics and climatic variability.

10-3-5   Multiple Cropping

Multiple cropping of cumin with onion is common in India and some parts of Iran. It seems there are possibilities for multiple cropping of cumin with other crops (e.g. peas). Diversification of low input production systems of cumin through multiple cropping is another research priority.

10-3-6   Response to Environmental Stresses

Cumin is usually grown in stress prone regions. However, its resistance to abiotic stresses such as drought, salinity, heat and cold has not been fully studied (7). Increasing cumin tolerance or resistance to environmental stresses is crucial for higher production.

10-3-7   Cumin Ideotype

So far cumin ideotype is not designed. It is not clear as to how extent traits such as branching, leaf size, or different yield components contribute to yield formation. Such questions could be answered by designing an ideotype.

10-3-8   Disease Control

Fusarium wilt and *Alternaria* blight are the major yield reducing factors in cumin production areas (1, 5). Despite several studies in different countries such as Iran and India, this problem still remains unresolved. To overcome this problem, identification of tolerant genotypes and development of new resistance cultivars is of great importance. Understanding interactions between cultivation practices and environmental conditions on disease development could also be important in control programs.

10-3-9   Production of Specific Herbicides

Cumin has very low resistance to weeds. On the other hand, low development rate of its leaf area leads to fast expansion of weeds. Mechanical weed control, mainly hand weeding, is the most common practice in production areas. This method is costly and could damage the crop as well. Introduction of selective herbicides for weed control in cumin fields should be set as a research priority. Such an attempt has already started but the results are uncertain (6, 8).

10-3-10   Mechanization

Cumin production systems are labor-dependent with low mechanization levels. While this system is important for employment of farmers, it reduces the uniformity and precision of cultivation practices. Development of specific machinery for cumin planting and harvesting is required for increasing economic return of producers.

10-3-11   Cumin Straws for Animal Feeding

Cumin straws are approximately equivalent to its grain yield. This straw, in addition to mineral nutrients, contains essential oils, which are effective in increasing the lactation rate of domestic animals. However, annually a huge amount of this valuable straw remains unused. More attention should be paid to the possibility of using cumin straws as an inexpensive source for animal feeding.

10-3-12   Essential Oils from Straw

Leaves and stems usually produce assimilated requirements of grains, therefore, the seed composition is influenced by the composition of vegetative organs. While seeds are the main source of cumin essential oils, the straws should also be considered as a potential source. Extraction of essential oils from cumin straws has been studied in Egypt with promising results (4).

10-4   CUMIN IN INTERNET

Cumin related Internet sites are mainly designed for the purpose of marketing or its application in food industries. The following sites contain useful scientific information about different aspects of this underutilized species:

http://www.hort.purdue.edu/newcropnexus/Cuminum-cyminum-nex.html

This site is developed by Purdue University and provides comprehensive information on new crops including cumin.

http://www.plantdatabase.com/go/282

In this site detailed information about world crop species is collected, with some general information on cumin.

http://www.ibiblio.org/pfaf/cgi-bin/arr-html?cuminum+cyminum

Information on cumin botany, grain physical characteristics, production areas, agronomy and its use in food industry as well as its medicinal properties can be obtained from this site.

10-5 SUMMARY

Despite the long history of cumin production in Iran and other Asian and African countries, because of its low cultivated areas, scientific research on this spice is scarce. Due to its commercial use in pharmaceutical companies and food and cosmetic industries of developed countries, existing scientific researches are mainly concentrated on cumin composition and properties.

Cumin production is seriously threatened by *Fusarium* wilt and *Alternaria* blight diseases. Weeds are also major yield reducing factors in cumin fields. In addition, different ecophysiological aspects of cumin growth and development, yield formation physiology, tolerance to abiotic stresses and its optimal agronomic practices are not fully understood. This gap of knowledge calls for multidisciplinary research programs in cumin producing countries. Universities and Agricultural Research Centers should put more emphasis on this valuable underutilized crop, and new research centers need to be established.

REFERENCES

1. Alavi, A. 1968. Cumin *Fusarium* wilt disease. Iranian Journal of Plant Disease 3: 92-98.

2. Aminpour, R. and S.F. Moosavi. 1995. Effects of irrigation intervals on developmental stages, yield and yield components of cumin. Agriculture and Natural Resources. 1: 1-7.

3. Chandula, R.P., S.C. Mathur and R.K. Srivastava. 1970. Cumin cultivation in Rajasthan. Indian Farming. July: 13-16.

4. El-Sawi, S.A. and M.A. Mohamed. 2002. Cumin herb as a new source of essential oil and its response to foliar spray with some micro-nutrients. Food Chemistry. 77: 75-80.

5. Hjian Shahri, M. 1996. Chemical control of cumin blight disease. Scientific and Technological Research Organization of Iran, Khorasan Center, Iran.

6. Kafi, M. and M.H. Rashed Mohassel. 1992. Effects of weed control intervals, row spacing and plant density on growth and yield of cumin. Agricultural Science and Technology. 6: 151-158.

7. Nabizadeh, M.R. 2001. Effects of different salinity levels on growth and yield of cumin. MSc. Thesis, Faculty of Agriculture, Ferdowsi University of Mashhad, Iran.

8. Rahimi, M. 1995. Evaluation of chemical weed control in cumin fields. Scientific and Technological Research Organization of Iran, Khorasan Center, Iran.

9. Sadeghi, B. 1991. Effects of nitrogen levels and irrigation on cumin production. Scientific and Technological Research Organization of Iran, Khorasan Center, Iran.

Index

About the Editors

Dr. Mohammad Kafi received his Doctorate from the University of Newcastle – upon-Tyne, UK. He presently works as an associate professor in the Department of Agronomy, Faculty of Agriculture, Ferdowsi University of Mashhad and is affiliated with the Center of Excellence for Special Crops (CESC). He has been teaching different courses on agronomy, crop production under abiotic stresses and crop physiology at undergraduate and graduate levels for the last sixteen years. Dr. Kafi has published more than 30 articles in international and national scientific journals and authored several books in Persian and English and also translated a number of foreign language books to the Persian language. The first research project that he undertook was his MSc. project on cumin and he continues his research on his plant till date. He also served as the Science Counselor at the Embassy of I.R. Iran in New Delhi, India (2003-2006) and during his tenure he visited many Indian universities and states that specialize in cumin research and production.

Dr. M.H. Rashed Mohassel graduated from University of Nebraska, USA. He is professor at the Department of Agronomy, Faculty of Agriculture, Ferdowsi University of Mashhad and is also affiliated with the Center of Excellence for Special Crops (CESC). He has been teaching different courses on weed sciences, botany and crop production at BSc, MSc and PhD levels for the last 25 years. Professor Rashed has published more than 70 articles in scientific journals and has written several books and has translated a number of books to the Persian language. He has been involved in several scientific associations at the national level. He is at present involved in MSc and PhD programs at the department and supervises theses on different aspects of weed management.

Dr. Alireza Koocheki did his PhD from the University of Wales, UK. A professor in the Department of Agronomy, Faculty of Agriculture, Ferdowsi University of Mashhad. Dr. Koocheki is also affiliated with the Center of Excellence for Special Crops (CESC). He has been teaching different courses on Production Ecology and Agroecology at BSc, MSc and PhD levels for the last 30 years. Professor Koocheki has published more than 100 articles in scientific journals and has authored/co-authored books and has translated a number of books to the Persian language. He has been involved in several consultation projects at the national level and has also taken part in a few international projects. He is at the present involved in supervising research projects of postgraduate scholars on different aspects of agricultural ecology and sustainable crop production in dry areas. He is also a fellow of Academy of Science of Islamic Republic of Iran.

Dr. Mahdi Nassiri graduated from the University of Wageningen, Netherlands. He is an associate professor and head of the Department of Agronomy, Faculty of Agriculture, Ferdowsi University of Mashhad and Director of the Center of Excellence for Special Crops (CESC). He has been teaching different courses on statistics in agriculture, ecology and agroecology and modeling at BSc, MSc and PhD levels for 15 years. Dr. Nassiri has published more than 20 articles in scientific journals and has authored several books. He has also translated a number of books to the Persian language.